The Economic Consequences of Climate Change

This work is published on the responsibility of the Secretary-General of the OECD. The opinions expressed and arguments employed herein do not necessarily reflect the official views of the Organisation or of the governments of its member countries.

This document and any map included herein are without prejudice to the status of or sovereignty over any territory, to the delimitation of international frontiers and boundaries and to the name of any territory, city or area.

Please cite this publication as:
OECD (2015), *The Economic Consequences of Climate Change*, OECD Publishing, Paris.
http://dx.doi.org/10.1787/9789264235410-en

ISBN 978-92-64-23540-3 (print)
ISBN 978-92-64-23541-0 (PDF)

The statistical data for Israel are supplied by and under the responsibility of the relevant Israeli authorities. The use of such data by the OECD is without prejudice to the status of the Golan Heights, East Jerusalem and Israeli settlements in the West Bank under the terms of international law.

Photo credits: © "sh12 on your wall" by Upupa4me, 2012 (CC BY-SA 2.0 *https://creativecommons.org/licenses/by-sa/2.0/*).

Corrigenda to OECD publications may be found on line at: *www.oecd.org/publishing/corrigenda*.

Preface

As we approach the Conference of the Parties (COP21) in late 2015 in Paris, our leaders are facing a fundamental dilemma: to get to grips with the risks of climate change or see their ability to limit this threat slip from their hands. Today our understanding of the scale of the risks posed by climate change is much better developed and supported by seriously tested and globally accepted evidence. This report, *The Economic Consequences of Climate Change* provides detailed insights into the consequences of policy inaction for the global economy. Its main contribution, when compared to previous efforts in this area, is that it does so through a more detailed economic modelling framework that links climate change impacts to sectoral and regional economic activity.

Trying to understand what climate change may mean for the future of our economies is daunting. It is not simply the case of coming up with a point estimate of what climate change might cost world Gross Domestic Product (GDP). What we need is a more nuanced understanding of how climate change impacts sectoral and regional economic activity, how these impacts propagate through our economic system, and what the downside risks are to long term economic growth. These insights, as provided by this report, will be invaluable in informing policy makers how to manage the significant and accumulating risk of serious climatic disruption.

The simulations carried out for this study suggest that in the absence of further action to tackle climate change, the combined negative effect on global annual GDP could be between 1.0% 3.3% by 2060. As temperatures could continue to rise to a projected 4°C above pre-industrial levels by 2100, GDP may be hurt by between 2% and 10% by the end of the century relative to the no-damage baseline scenario. Most importantly, the net economic consequences would be negative in 23 of the 25 regions modelled in the analysis, and particularly severe in Africa and Asia, where the regional economies are vulnerable to a range of different climate impacts.

The analysis in this report is not a prediction of what will happen, nor a synthesis of all the social costs of climate change. There is still a lot we cannot quantify, particularly with regard to the economic consequences of triggering important tipping points in the climate system which could be catastrophic for our economies. However, just like the build-up of risks before the financial crisis, uncertainty should not be an excuse for inaction. This report also demonstrates how early and ambitious action on adaptation and mitigation can significantly reduce these downside risks.

The OECD will continue to assist member and partner countries as both the challenges from climate change and the imperative to take stronger and more decisive action become more evident.

Angel Gurría
OECD Secretary-General

Acknowledgements

This report was prepared by Rob Dellink and Elisa Lanzi of the OECD Environment Directorate. Matthias Kimmel (OECD) provided substantial inputs to the drafting, and Jean Chateau (OECD) contributed to the modelling of ENV-Linkages. Kelly de Bruin (CERE, Sweden) was the lead modeller for the AD-DICE model simulations. Francesco Bosello and Ramiro Parrado (CMCC, Italy) provided expert guidance on the incorporation of climate change impacts in economic models and delivered the input information for several climate change impacts included in the modelling, not least fisheries, sea level rise, diseases and tourism. Substantive inputs for specific elements of the analysis were received from Yasuaki Hijioka (NIES, Japan), Yasushi Honda (University of Tsukuba, Japan), Juan-Carlos Ciscar and Zoi Vrontisi (IPTS-JRC, EU), Hessel Winsemius (Deltares, The Netherlands), Philip Ward (VU Amsterdam, The Netherlands), Petr Havlík (IIASA, Austria) and Ian Sue Wing (Boston University, USA). These collaborators are not responsible for the final report.

The OECD Environment Policy Committee (EPOC) was responsible for the oversight of the development of the report. In addition, the Working Party on Climate Change, Investment and Development (WPCID), Working Party 1 of the Economics Policy Committee (EPC-WP1) and the experts following the CIRCLE project reviewed earlier drafts.

The project was managed by Shardul Agrawala, who also provided feedback on the modelling and earlier drafts. Marie-Jeanne Gaffard provided administrative and technical support; statistical and technical assistance was provided by François Chantret. Janine Treves and Katherine Kraig-Ernandes supported the publication process of the final manuscript. This final version also benefits from comments on an earlier draft by Nils-Axel Braathen, Simon Buckle, Anthony Cox, Guillaume Gruère, Ada Ignaciuk, Giuseppe Nicolletti, Andrew Prag, Simon Upton (all OECD), as well as Juan-Carlos Ciscar (JRC-IPTS, Spain), Sam Fankhauser (LSE-GRI, UK), Tom Hertel (Purdue University, USA), Anil Markandya (BC3, Spain), Paul Watkiss (UK) and one anonymous reviewer.

Table of contents

Figures

Acronyms and abbreviations

CES	Constant elasticity of substitution
CGE	Computable general equilibrium
CO_2	Carbon dioxide
ECS	Equilibrium climate sensitivity
EU	European Union
GBP	Pound Sterling
GDP	Gross domestic product
GHG	Greenhouse gas
Gt	Gigatonnes
IAM	Integrated assessment model
IPCC	Intergovernmental Panel on Climate Change
NIES	National Institute for Environmental Studies
OECD	Organisation of Economic and Co-operation and Development
PPM	Parts per million
PPP	Purchasing power parities
PRTP	Pure Rate of Time Preference
RCP	Representative concentration pathway
SRES	Special Report on Emissions Scenarios
TFP	Total factor productivity
TOE	Tons of oil equivalent
UK	United Kingdom
UN	United Nations
US	United States
USD	United States dollars
VOLY	Value of a life year
VSL	Value of a statistical life
WTP	Willingness to pay

Executive summary

This report provides a detailed global quantitative assessment of the direct and indirect economic consequences of climate change (i.e. climate damages) for a selected number of impacts: changes in crop yields, loss of land and capital due to sea level rise, changes in fisheries catches, capital damages from hurricanes, labour productivity changes and changes in healthcare expenditures from diseases and heat stress, changes in tourism flows, and changes in energy demand for cooling and heating. Other major impacts of climate change are investigated outside the modelling framework.

Most existing studies of climate damages have a stylised, aggregated representation of the economy. This report uses a detailed multi-sectoral, multi-regional dynamic general equilibrium modelling framework (the OECD ENV-Linkages model) to link climate change impacts to specific aspects of economic activity, such as labour productivity, the supply of production factors such as capital, and changes in the structure of demand. This detailed analysis is used to assess damages until 2060, and is complemented by a more stylised analysis of post-2060 damages (using the AD-DICE integrated assessment model).

This report presents only one possible economic scenario and it cannot capture all impacts of climate change. It is not a prediction of what will happen, nor a synthesis of the full consequences of climate change. It sheds light, however, on how the selected impacts affect the composition of Gross Domestic Product (GDP) over time and how sectoral consequences spill over to other sectors and regions.

The modelling is based on existing estimates of how selected climate impacts affect the drivers of economic growth of major world regions at the macroeconomic and sectoral level. This so-called production function approach allows for a detailed assessment of a subset of the direct and indirect consequences of climate change for the economy for a selected number of climate change impacts. The analysis assumes no mitigation actions are taken beyond those that are already adopted, and only market-driven adaptation measures are considered.

The projections do not capture all considerable uncertainties and risks from climate change that could potentially lead to much larger damages (especially in the long-run), or result in smaller economic consequences than in the central projection. Some major uncertainties stem from assumptions on economic growth, demographics, the response of the climate system to increasing greenhouse gas (GHG) concentrations, projections of regional climate, and the valuation of climate change impacts. Large downside risks of climate change are associated with uncertainty about the response of the climate system to temperature increases beyond 2°C, irreversible tipping points, and the non-market impacts of climate change. Of these sources of uncertainties, the range of GDP consequences provided below reflects only the uncertainties relative to the response of the climate system to increasing GHG concentrations.

Key findings

- The ENV-Linkages model simulations suggest that market damages from the selected set of impacts are projected to gradually increase over time and rise faster than global economic activity. **If no further climate change action will be undertaken, the combined effect of the selected impacts on global annual GDP are projected to rise over time to likely levels of 1.0% to 3.3% by 2060**, with a central projection of 2%. This range reflects uncertainty in the equilibrium climate sensitivity (ECS) – a measure indicating how sensitive the earth's climate reacts to a doubling of atmospheric CO_2 – using a likely range of 1.5°C to 4.5°C. Assuming a wider range of 1°C to 6°C in the ECS, GDP losses could amount to 0.6% to 4.4% in 2060.
- As temperatures continue to rise to a projected 4°C above pre-industrial levels by 2100, AD-DICE projections suggest that **GDP may be hurt by between 2% and 10% by the end of the century relative to the no-damage baseline scenario** (under the likely ECS range). As experimental projections with the AD-DICE model show, continuing to emit greenhouse emissions as usual until 2060 will commit the world to economic damages in a range of 1% to 6% by the end of the century even if emissions fall to zero in 2060. However, assessments of impacts for higher temperature increases are much less robust; they could even lead to damages of 12% by 2100 when non-linearities in the climate damage function are strong.

Of the impacts modelled in the analysis, changes in crop yields and in labour productivity are projected to have the largest negative consequences, causing loss to annual global GDP of 0.9% and 0.8%, respectively, by 2060 for the central projection. Including a CO_2 fertilisation effect reduces the agricultural damages to 0.6%, and the effect is projected to be especially strong in Africa (reducing agricultural damages from 1.5% to 1.0% by 2060 in Sub-Saharan Africa).

Damages from sea level rise also gradually become more important, growing most rapidly after the middle of the century. Damages to energy and tourism are very small from a global perspective, as benefits in some regions balance damages in others. Climate-induced damages from hurricanes may have significant effects on local communities, but the macroeconomic consequences are projected to be very small.

- **Net economic consequences are projected to be negative in 23 of the 25 regions modelled in the analysis. They are especially large in Africa and Asia, where the regional economies are vulnerable to a range of different climate impacts, such as heat stress and crop yield losses.** GDP losses in 2060 for the selected impacts covered in the analysis are projected to amount to 1.6% to 5.2% in the Middle East and North Africa regions, 1.7% to 6.6% in the South- and South-East Asia regions (including India) and 1.9% to 5.9% in the Sub-Saharan Africa regions, respectively (using Purchasing Power Parities exchange rates to aggregate across regions). Again, these regional projections only take into account uncertainty from equilibrium climate sensitivity; moreover, uncertainties are larger on the regional than on the global level.
- **The model results show that for some countries, especially those in higher latitudes, the beneficial economic consequences of the impacts considered in the analysis are projected to outweigh the damages from climate change, at least to 2060.** Economic benefits stem predominantly from gains in tourism, energy and health. The global assessment also shows that countries that are relatively less affected by climate change may reap trade gains. These projections do not, however, include potential negative

effects from the occurrence of climatic tipping points as well as other climate change impacts not modelled in the assessment. Local effects may also differ significantly from the national averages.

- **The actual magnitude of the regional damages will depend in part on the ability of economies to adapt to climate impacts by changing production technologies, consumption patterns and international trade patterns.** For instance, reductions in availability of land and capital due to sea level rise are projected to induce a reallocation of land and capital between sectors and thus affect the entire economy. The significance of indirect effects on sectors and regions confirms the importance of using a multi-sectoral, multi-regional economic approach. For more severely affected countries, especially India, there are also non-negligible interaction effects, and the total GDP loss is smaller than the sum of the individual losses from different impacts, indicating that countries can respond to the variety of different impacts in a more sophisticated way than simply responding to each individual impact separately.
- The modelling approach applied in this study can only provide a partial picture of the consequences of climate change, as it cannot take into account non-market aspects of well-being (e.g. premature deaths) or impacts for which the available data are insufficient.
  - **Urban flood damages** are highly uncertain, in part because they rely on projections of regional and local precipitation, as well as behavioural responses. Moreover, only potential damages in absence of adaptation efforts could be assessed. The two countries that have by far the largest projected potential urban flood damages are India and China; for OECD countries, the climate-induced potential urban flood damages are projected to be much smaller.
  - The regions with the highest number of **premature fatalities from heat stress** are the ones with high population (like China and India) or where aging increases the size of the vulnerable population at risk (such as the EU and the US).
  - For **loss of ecosystem services**, a value of around 1% of GDP by 2060 is projected for most high-income countries, based on an approach to calculate the Willingness-to-Pay for protection of these services.
  - While the temperature thresholds associated with the **triggering of high-impact large-scale singular events** in the climate system remain uncertain, in general terms their likelihood increases with more severe climatic changes, and they are expected to have severe permanent effects on the economy.

On balance, the costs of inaction presented here likely underestimate the full costs of climate change impacts. Without attempting to be complete, the report also qualitatively discusses a number of important climate impacts that could not be quantified, including impacts on reduced winter mortality from extreme cold, local disruptions of infrastructure from extreme weather events, changes in water stress and impacts on human security (specifically migration and conflict). Although for some of these effects, and for particular regions, the consequences may be positive, the existing evidence collated by the Intergovernmental Panel on Climate Change (IPCC) and others points to significant further downside risks for negative consequences.

Key policy conclusions

The economic consequences of climate change, as outlined in detail in the report, with losses in GDP for almost all regions and numerous important other consequences, imply a strong call for policy action. By implementing ambitious mitigation policies to reduce the emission sources of climate change, and adaptation policies to best deal with the remaining consequences, the worst impacts and risks may be avoided, and the economic consequences from climate change substantially reduced.

- Besides projected market damages, policy makers need to take into account the large downside risks and long-term effects of climate change when designing mitigation and adaptation policies. Emission reductions lead to a stream of future benefits and reduced risks, while adaptation reduces the adverse consequences of climate impacts that are already underway and helps societies proactively prepare for the future. Therefore, the calculation of the benefits of policy action should be based on the full stream of future avoided impacts, and not simply follow the time profile of market damages as they emerge.
- The benefits of adaptation policies, from a reduction in the selected damages alone, may amount to more than 1 percentage points of GDP by the end of the century, as the stylised analysis with AD-DICE shows. It also highlights that if barriers to adaptation are strong, and firms and households are not able to adapt at all, the costs of climate change can even double.
- Early and ambitious mitigation action (aimed at minimising total climate costs over the long term) can help economies avoid half of the damages to GDP by 2060; calculations of avoided damages exclude the economic effects of the mitigation policy itself. It can also reduce the risk of being locked into the negative long-term consequences of climate change. Despite the capacity of mitigation to limit impacts, however, significant damages from climate change are projected to persist in vulnerable regions, such as in most countries in Africa and Asia.
- Mitigation not only reduces the expected level of climate damages, but ambitious mitigation action also considerably reduces the risks of high damages (the likely uncertainty range reduces damages from 2-10% to 1-3% by 2100 for the selected climate impacts, according to the simulations). Furthermore, less ambitious mitigation policies in the first decades will have lower short-term costs, but lead to higher long-term risks (in quantitative terms, this result is heavily influenced by the choice of discount rate).
- Mitigation policies will reduce the negative impacts of climate change on all economic sectors, yet the costs of these policies will not be borne by all sectors proportionally to their expected benefits. Both damages and the mitigation policy lead to a shift in the structure of the economy towards more services. The detailed economic modelling analysis is used to shed further light on this, again with a horizon to 2060.
  - **Agriculture**, for example, despite its relatively small size, will experience substantial direct and indirect impacts from climate damages; its high emissions could imply substantial costs from stringent economy-wide mitigation policies.
  - For **energy production and the industrial sectors** the climate damages are smaller than the potential effects from stringent economy-wide mitigation policies. Renewable power generation can substantially increase production activities if an ambitious mitigation policy is implemented, but on balance the negative effects on fossil fuel producers outweigh those on renewables.
  - **Services** are projected to benefit from the mitigation policy as they are relatively clean, but they are negatively affected by climate damages. However, given the large size of services compared to the other sectors, the relative share of the services sectors in total GDP can increase, i.e. they are *relatively* less affected than other sectors.

Additional research efforts are warranted to reduce the major knowledge gaps on the damages of climate change, not least concerning the regional economic consequences of triggering important tipping points, which could potentially have effects on the economy that are an order of magnitude higher than those included in the modelling analysis here. Furthermore, a robust methodology is needed to include non-market damages and co-benefits of policy action into the evaluation.

An optimal climate policy mix will require both adaptation and mitigation, while also including a certain level of residual damages. When policy makers assess the costs and benefits of policy action, they need to think of including a risk premium to reflect the risks of crossing irreversible tipping points, and to avoid the downside risks of more severe damages. Finally, there are important co-benefits from some policy actions that can be reaped immediately and locally, such as air quality benefits. Policy makers need to take these into account when determining the appropriate policy efforts.

Chapter 1

Modelling the economic consequences of climate change

This chapter first presents a brief discussion of the main categories of climate change impacts. It then introduces the methodology used to identify how climate impacts affect economic growth. It highlights how the costs of inaction until 2060 can be assessed using a production function approach to link climate change impacts to specific drivers of growth in the dynamic general equilibrium model ENV-Linkages; and how and why this is complemented by a more stylised integrated assessment modelling of long-term impacts using the AD-DICE model. The chapter ends with a description of how the production function approach is used to model the various impacts in ENV-Linkages.

1.1. Introduction

Evidence is growing that changes in the climate system are contributing to a range of biophysical and economic impacts that are already affecting the economy (e.g. see the latest reports of the Intergovernmental Panel on Climate Change: IPCC, 2013, 2014a,b; see also Dell et al., 2009, 2013). Future impacts are expected to be much larger (IPCC, 2014a). A certain amount of climate change is already locked-in, and there are considerable and cascading uncertainties with regard to future emissions of greenhouse gases, the resulting changes in climate, and the resulting biophysical and socioeconomic impacts. It can therefore reasonably be asked what value a modelling analysis of the economic consequences of climate change at a global level can offer policy makers. After all, a combination of these uncertainties and the necessary simplifications of any model representation of the global economy mean that the absolute magnitude of one point estimate of a specific impact of climate change on the economy will be less interesting than the interactions in the economic system that they induce.

The particular value of this exercise is to produce a carefully caveated account of the costs of not mitigating the global emissions trajectory and, conversely, the benefits associated with action. Policies aiming to limit climate change impacts will have global economic consequences (even if the policies are not applied globally). As its name suggests, the ENV-linkages model is designed to shed light on *linkages* – on the way physical impacts can affect patterns of production leading to changes in the composition of growth regionally and sectorally. While the magnitude and distribution of these changes is uncertain, modelling provides some clues to long-run trends and mechanisms that link climate change impacts and economic activity. These at least should be valuable in informing attempts to manage the significant and accumulating risk of serious climatic disruption.

The report presents results from modelling the feedbacks of climate change damages on economic growth for the coming decades. It paints a picture of a world in which a dynamic global economy internalises the damages of climate change. In the process it continues to deliver huge gains in global output and, unevenly, living standards. What it cannot tell us is how fragile or leveraged those living standards will be given the increasing risk of costly extreme events and non-linear change that accompany that growth process.

The impacts of climate change will play out over a very long time period. Given the large uncertainties in projecting its course and impacts, the costs of inaction cannot be subjected to a simple cost benefit analysis. Rather, any comparison of the costs and benefits of different policy mixes needs to be based on an assessment of risks that incorporates the inter-temporal dimension of the problem. For this reason, the report goes no further than providing a stylised assessment of some of the main benefits of policy action, including both mitigation and adaptation policies is included.

There is extensive literature on the economic impacts of climate change (e.g. Nordhaus, 1994, 2007, 2010; Tol, 2005; Stern, 2007; Agrawala et al., 2011) and on modelling the costs of policy action (e.g. OECD, 2012). In-depth regional studies on the consequences of climate change also exist, most notably the Garnaut Review for Australia (Garnaut, 2008, 2011), the Risky Business study for the USA (Risky Business Project, 2014), the Peseta project for the European Union (Ciscar et al., 2011, 2014) and the COIN study for Austria (Steininger et al., 2015). Some literature has also attempted to quantify the costs of inaction and benefits of policy action on climate change. Most notably, the Stern Review (2007) concludes that climate change could reduce welfare by an amount equivalent to a *permanent* reduction in consumption per capita of between 5% and 20%. The size of the effects of climate impacts on the economy is, however, still the subject of debate, as confirmed by Working Group II of the IPCC (2014a) which concludes that economic impact estimates produced over the past in the past two decades "vary in their coverage of subsets of economic sectors and depend on a large number of assumptions, many of which are disputable, and many estimates do not account for catastrophic changes, tipping points, and many other factors. With these recognized limitations, the incomplete estimates of global annual economic losses for additional temperature increases of ~2°C are between 0.2 and 2.0% of income (± 1 standard deviation around the mean) (medium evidence, medium agreement)" (IPCC, 2014a).

Most of these studies have a stylised, aggregated representation of the economy. Typical modelling studies that focus on projections of climate change impacts over time include highly aggregated Integrated Assessment Models (IAMs), in which climate damages in different sectors are aggregated and used to re-evaluate welfare in the presence of climate change. Comparing such models is difficult, as each tends to include different impact categories, but it is clear that they vary widely in their projections for the global macroeconomic consequences for specific impacts (e.g. US Interagency Working Group, 2010; 2013). A much smaller strand of literature uses computable general equilibrium (CGE) models to examine the economic implications of climate change impacts in specific sectors, often using a comparative static approach (e.g. Bosello et al., 2006; 2007). Box 1.1 briefly introduces the main differences between these types of models. More recently, CGE models have also been used to study the economy-wide impacts of climate change in a dynamic setting (see Eboli et al., 2010; Bosello et al., 2012; Roson and Van der Mensbrugghe, 2012; Bosello and Parrado, 2014; Dellink et al., 2014).

Box 1.1. **Computable general equilibrium and integrated asessment models**

The two main types of models used for assessing the economic consequences of climate change are Computable General Equilibrium (CGE) models and Integrated Assessment models (IAM).

CGE models focus on the relations between different economic actors and contain a full description of the economic system using multiple economic sectors: households supply production factors (labour, capital, land) and consume goods and services, while firms transform the production factors, with intermediate deliveries from other sectors, into the output of goods and services. In a multi-regional CGE model all economies are linked through international trade. For assessing climate change damages, the detailed description of the economy allows for a detailed representation of those impacts of climate change that are primarily affecting markets, such as changes in crop yields, health expenditures, labour productivity and energy demand.

Box 1.1. **Computable general equilibrium and integrated asessment models** *(cont.)*

IAMs focus more on describing the interactions between the economic and biophysical system, i.e. how economic activity leads to environmental pressure, and how environmental feedbacks affect the economy. Many IAMS that have been used for policy advice (such as DICE, FUND and PAGE) are highly aggregated and contain only a cursory description of the economy. Other IAMs have much greater detail in the description of the biophysical system, often at the expense of lacking feedbacks from the biophysical system to the economy. The more stylised nature of IAMs make them more suited to describe a wider range of climate change impacts in an aggregated fashion.

In principle, there is no clear distinction between both types of models: enhanced CGE models such as ENV-Linkages describe emissions from economic activity in detail and contain feedbacks from the climate impacts on the economic system, and is thus de facto an IAM. Similarly, the economic module of an IAM can be expanded into a full-fledged CGE model. The level of detail that can be captured in CGE models and IAMs is limited by computing power and, more importantly, the need to avoid the model becoming so complex that it is a black box.

The specification of climate change impacts in CGEs and IAMs is further discussed in Section 1.3.

This report builds on these recent studies and it uses the OECD's multi-region, multi-sector dynamic CGE model ENV-Linkages to analyse the economic consequences of a selection of climate change impacts until 2060. By using a detailed economic model, with explicit representation of the drivers of economic activity, the impacts of climate change can be linked to the economy in a much more realistic fashion. The analysis with the ENV-Linkages model is complemented with an assessment of consequences of climate change after 2060 and for a stylised analysis of the benefits of policy action with the integrated assessment model AD-DICE. While the sectoral and regional details of the CGE model are ideal for a detailed study of the consequences of climate change on the various parts of the global economy, and especially the wider economic consequences that trickle through the economy, the optimisation structure of the IAM model is better suited to study the policy trade-offs and longer term consequences. Both models use the same baseline scenario for socioeconomic developments (including population and GDP). For OECD countries and the main emerging economies, this is based on the OECD long-run aggregate growth scenario to 2060 (OECD, 2014a).

Chapter 2 of this report focuses on how a selected number of climate change impacts affect different parts of the economy. The impact categories that are investigated in Chapter 2 include some of the major impacts with respect to agriculture, coastal zones, extreme events, health, energy and tourism demand. For most impact categories, a number of the key economic impacts are included in the modelling exercise, while non-market impacts are discussed separately in Chapter 3. For other impacts, including those related to ecosystem services, water stress and tipping points, only anecdotal evidence is presented, as sufficiently robust data to study economic damages is not available.

In order to provide an indication of magnitude in a metric that is widely known to policymakers, the resulting impacts from both models are presented in terms of effects on gross domestic product (GDP). This is an imperfect measure of the total economic costs of climate change, since it does not consider wider consequences on well-being or costs to society (which can be considerable). Nevertheless, it provides insight in the macroeconomic consequences (i.e. economic feedbacks) of the selected climate change impacts; the sectoral decomposition of GDP can further illuminate the changes in economic structures associated with climate change damages. Moreover, expressing these costs of inaction in the same terms as the usual indicator for economic growth, i.e. in terms of GDP losses, helps to communicate the importance of climate change for mainstream economic policy making.

1.2. Main consequences of climate change

Climate change will have pervasive socio-economic consequences that will not only affect major economic sectors such as agriculture, energy or healthcare, but will also result in changes to the supply and demand for goods and services of all sectors of the economy, albeit with varying levels of intensity. Higher temperatures, sea level rise, and other climatic changes (changes in regional precipitation patterns, the water cycle, frequency and intensity of extreme weather events), will also impact aspects of life that are not primarily based on or related to economic activity, as for example human security, health and well-being, culture, people's capabilities, and environmental quality.

Table 1.1 provides an overview of the selection of climate impacts considered in this report. It is important to note that these are a subset of all impacts of climate change, even for the sectors covered. Not all of these impact categories are entirely discrete nor can they always be clearly separated from each other. Extreme events, for example, not only affect human health, land and capital damages, but might induce people to migrate to other places as a form of adaptation; extreme events can also cause long-lasting trauma for those directly or indirectly affected by their long-run consequences. Agriculture is highly dependent on

Table 1.1. **Categories of climate impacts considered in this study**

AGRICULTURE	Changes in crop yields (incl. cropland productivity and water stress)	Modelled
	Livestock mortality and morbidity from heat and cold exposure	Qualitatively
	Changes in pasture- and rangeland productivity	Stand-alone
	Changes in aquaculture productivity	Qualitatively
	Changes in fisheries catches	Modelled
COASTAL ZONES	Loss of land and capital from sea level rise	Modelled
	Non-market impacts in coastal zones	Qualitatively
EXTREME EVENTS	Mortality, land and capital damages from hurricanes	Modelled
	Mortality, land and capital damages from floods	Stand-alone
HEALTH	Mortality from heat exposure (incl. heatwaves)	Stand-alone
	Morbidity from heat and cold exposure (incl. heatwaves)	Modelled
	Mortality and morbidity from infectious diseases, cardiovascular and respiratory diseases	Modelled
ENERGY DEMAND	Changes in energy demand for cooling and heating	Modelled
TOURISM DEMAND	Changes in tourism flows and services	Modelled
ECOSYSTEMS	Loss of ecosystems and biodiversity	Stand-alone
	Changes in forest plantation yields	Qualitatively
WATER STRESS	Changes in energy supply	Qualitatively
	Changes in availability of drinking water to end users (incl. households)	Qualitatively
HUMAN SECURITY	Civil conflict	Qualitatively
	Human migration	Qualitatively
TIPPING POINTS	Large scale disruptive events	Stand-alone

Note: "Modelled" implies that the impact is captured (at least partially) in the main modelling framework; "stand-alone" refers to a quantitative assessment outside the main modelling framework, and "qualitatively" implies only a qualitative assessment was possible in this report.
Source: Own compilation.

functioning ecosystems and water availability, while damages in coastal zones affect, inter alia, ecosystems, livelihood, and agriculture. To avoid double-counting, all impacts considered in this study are allocated to only one impact category, along the lines of Table 1.1. Table 1.1 also indicates whether these impacts are included in the modelling exercise, are part of a stand-alone quantitative assessment, or are discussed qualitatively.

Working Group II of the IPCC (IPCC, 2014a) describes the most significant projected impacts of climate change to affect the economy, society, and the earth's environment in various scenarios (including greenhouse gas concentration pathways), where possible attaching information about the level of likelihood, evidence, and agreement on the findings or relationships between climate change and the impacted variables. It is not the aim of this report to summarise all possible impacts. Rather, without aiming to be complete, the following paragraphs give some selected examples of important impacts of climate change that may occur within the categories presented in Table 1.1. Many other impacts have been assessed by the IPCC, but could not be included in the assessment of this report.

In *agriculture*, climate change will have consequences for various subsectors, including crop production, livestock, pasture- and rangeland, and aquaculture. Of the various climatic drivers, the impacts of climate change (including changes in regional temperatures and precipitation patterns) on crop productivity have been studied most comprehensively, suggesting that at the global level the impacts will be largely negative for moderate to high levels of warming (Rosenzweig et al., 2013). At the regional level, however, there will be large differences among regions with positive impacts in some and negative impacts in others. Changes in rainfall, atmospheric carbon dioxide (CO_2) and ozone concentrations, changes in pest and disease prevalence, and extreme events spurred by climate change will likely also affect future agricultural activities, sometimes positively and sometimes negatively. According to the IPCC (2014a), there is high confidence that higher CO_2 concentration in the atmosphere will have a stimulatory effect on crop yields (but also on weeds), while higher levels of ozone are likely going to be damaging. In addition, climate change might have consequences for outcomes that depend on the way agriculture is conducted, including conservation of the countryside, food security and the maintenance of biodiversity. It might also affect negative externalities produced by agricultural activity, as for example soil and water pollution. The evidence on this relationship and its direction is not very clear, however (Ahlheim and Frör, 2003; OECD, 2001). OECD (2014b) discusses in detail the effects of climate change on the agricultural sector through the water system. With regards to fisheries, climate change is expected to negatively impact most developing countries – especially those located in tropical regions. Developed nations at more northern latitudes, in turn, may benefit (IPCC, 2014b).

Coastal zones or coastal systems include natural ecosystems (beaches, cliffs, lagoons, etc.) and human systems (settlements, cities, ports, food production, etc.). They comprise distinct coastal features and ecosystems, as well as built environment, human activities, and institutions that organize these human activities (IPCC, 2014a). Various climate change-related drivers can impact on these systems. Beyond likely changes in the frequency and intensity of storms (and storm surges), increases in precipitation, warmer ocean temperatures and ocean acidification, sea level rise is potentially the most significant contributing factor to coastal zone damage. There is high agreement among authors of the IPCC and other reports that rising sea levels can negatively impact the provision of market and non-market goods and services in coastal zones through events such as storm surge, submergence, salt-water intrusion and coastal erosion. Both natural and human systems will be affected by sea level rise.

Extreme weather events are very likely to be affected by climate change, although the regional changes vary. Tropical cyclone activity, including hurricanes, is more likely than not to become more intense by the end of the 21st century, as global mean surface temperature rise (IPCC, 2014a). The IPCC (2013) reports that with higher temperatures, extreme precipitation events over most of the mid-latitude land masses and over wet tropical regions will very likely become more intense and more frequent by the end of this century. Similarly, river floods are also projected to increase in number and severity in most river basins. Reduced rainfall and increased evaporation can both lead to droughts, which are projected to "become longer, or more frequent, or both, in some regions and seasons" (IPCC, 2014a). These trends and their related damages are projected to result in higher costs to the economy relative to a world without climate change. The main direct channels through which economies will experience these damages are impacts on physical capital (e.g. factories, houses, streets and bridges, machinery, computers, but also energy infrastructure), land (e.g. natural resources), and labour (i.e. the workforce). These events also lead to indirect economic effects, e.g. through the disruption to electricity supply or transport, or a temporary halt to almost all local economic activity. The increase in frequency and intensity of extreme events as a consequence of climate change also leads to premature deaths and injuries, and force people to leave their homes and temporarily or permanently move to other places, affecting well-bring and welfare. They also impact on ecosystems and the services these provide. Evidence is also emerging that economies do not fully recover from the macroeconomic costs of destruction but are permanently faced with lower levels of GDP and economic growth (Hsiang and Jina, 2014), although this may depend on the level of development and the stock of physical and human capital. Logically, this extends to climate-induced destruction.

Health impacts of climate change include both direct and indirect effects, including: heat and cold related mortality and morbidity, water, food and vector-borne disease; deaths and well-being; and changes in air pollution and allergens. There are also risks to health infrastructure and to occupational health (WHO, 2012; 2014). The economic costs of health impacts are not easy to assess as they include both market and non-market costs. For instance morbidity costs include market impacts, such as the effects of illnesses on labour productivity, and non-market impacts, such as the costs of pain and suffering.

The *demand for energy* will also be affected by climate change. The main channels for changes in energy demand are through reduced need for heating in winter, and increased need for cooling in summer. Energy supply may also disrupted, e.g. by water shortages, and this may in turn affect energy demand. The projected changes in the energy system are, however, dominated by the assumptions on mitigation policies. IEA (2013) investigates the links between climate change and the energy system in detail.

With regards to *tourism*, the effects of climate change arise from changes in local climate conditions, making certain tourist locations less attractive and others more. For instance, skiing in the Alps may become less snow-secure, and the high cost of providing artificial snow increases prices for Alpine skiing. This induces changes in both domestic and international tourist flows, plus changes in their expenditures.

Ecosystems on land and in water provide a multitude of precious services to humans and other species, including the supply of food, raw materials, climate and air quality, habitat for species, and opportunity for aesthetic appreciation and inspiration (TEEB, 2014). Climate change is expected to place ecosystem services under further stress – directly as well as indirectly by interacting with and intensifying other aggravating factors, such as human development. Warming, as a major direct climatic driver, and changes in extreme events, will likely reduce biodiversity and diminish abundance of species, or – if possible – force certain species (both animals and plants) to shift range to higher latitudes or higher elevations with more bearable temperatures to increase the chance of survival. Northward migration of fish and birds (and tree species in general) is one example for range shift as a response to warming in the Northern Hemisphere. Heavy precipitation, in turn, might act as an indirect impact on ecosystems, by accelerating the erosion of forest areas that have already been put under pressure, e.g. from recent logging (EPA, 2015). Changes in availability and quality of ecosystem services will also affect the functioning of economic sectors, not least the land-based agricultural sector, forestry and fisheries.

Climate change is projected to have both positive and negative impacts on *freshwater resources*, with the effect varying to a large extent by geographic latitude. While the global circulation climate system models vary significantly in their projections of regional climate changes, including precipitation patterns, it is expected that many humid mid-latitude and high latitude regions will most likely experience increased water availability with climate change. Groundwater is the biggest reservoir of available freshwater and is relatively better insulated from climate change. Nonetheless, groundwater recharge is projected to decline in many countries and sea level rise may increase salinity of groundwater reservoirs. Declining water availability and a larger number of extended dry periods are projected to affect drier many countries in the mid-latitudes and dry subtropical latitudes, although uncertainties on regional water availability are very large. Short-term or seasonal water reductions from more variable streamflow (mostly resulting from a greater variability in rainfall) and reduced storage of water in ice and snow might nonetheless also be felt in regions with projected larger water availability. In addition, negative impacts of climate change on water quality from toxins produced by algae, for example, can contribute to reduced availability of freshwater (OECD, 2012, 2013; IPCC, 2014a). These impacts are expected to affect, inter alia, end users through changes to the availability of drinking water as well as industry through impacts on water supply for irrigation and energy supply.

According to the IPCC (2014a), there is high confidence and robust evidence that climate change will intensify stressors that negatively impact *human security*, which can be defined as "a condition that exists when the vital core of human lives is protected, and when people have the freedom and capacity to live with dignity" (IPCC, 2014a). Forced migration and incidence of civil conflict are two key stressors to human security that have been widely discussed in the literature and that many expect to be magnified by climate change. However, evidence for direct causal linkage between climate change and these specific factors is still limited, and the linkage itself is contested by some.

Besides the changes occurring in the various sectors and regions as described above, there is a risk associated with large-scale disruptions caused by climate change (so-called large singular events). These large-scale events, or *tipping points* (tipping elements), can occur when small climate changes trigger a disproportionately large impact and thus pose a systemic risk. Models cannot easily assess the implications of major climate events, such as a collapse of North Atlantic thermohaline circulation (i.e. shut-down of the Gulf Stream) or abrupt solid ice discharge of the West Antarctic ice sheet. While most large-scale singular events are unlikely to occur in the 21st century (IPCC, 2013), with the exception of the partial loss of artic sea ice, the risks associated with the potential for a large and irreversible sea level rise from ice sheet loss "increase disproportionately as temperature increases between 1-2°C additional warming and become high above 3°C" (IPCC, 2014a). These risks potentially have very large consequences for the world economy, yet the changes in the climate system that trigger them – and the thresholds when these events may occur – are poorly understood, and the economic consequences de facto impossible to robustly project.

While there is mounting evidence that there are significant downside risks from large singular efents and other climate impacts, insofar these are related to different uncertainties and are thus largely independent, there is only a very small chance of all of them occurring. It is more likely that some of these risks may occur while other do not materialise. But when these risks are positively correlated (which is the case for those that are related to global temperature increases), then these risks may well combine and their probabilities move together.

These impact categories listed in Table 1.1 are used throughout the report to describe the methodology and the results of the analysis. As described in Section 1.4, several of these impacts are modelled in ENV-Linkages to estimate the costs of climate change inaction to 2060. That does not mean that the other impacts do not have economic consequences, but that there is not enough information to include them in the model, or that the impacts primarily have non-market consequences that cannot be readily included in an economic modelling framework. Such impacts are discussed in Chapter 3.

1.3. A framework to study climate change impacts on economic growth

1.3.1. *A multi-model framework*

A standard framework to assess climate damages begins by linking economic activities to emissions of greenhouse gases (GHG), and evaluating how human-induced increases in atmospheric GHG concentrations drive changes in climate, such as changes in regional and global temperature and precipitation patterns. These climatic changes in turn result in physical and biogeochemical impacts which influence the productivity of various sectors of the regional economies where the impacts occur, and ultimately give rise to economic losses.

This approach combines representations of some or all of the following components: the determinants of socioeconomic development, emissions caused by economic growth, the atmosphere-ocean-climate system, ecosystems, socioeconomic impacts, mitigation and adaptation policies and associated economic responses, with different types of models emphasizing different linkages (Parsons and Fisher-Vanden, 1997).

This report combines two models in one complementary framework in order to capture as many aspects as possible. A sectoral and regional computable general equilibrium model is used wherever possible, and the analysis is combined with that using a large scale integrated assessment model when needed. While the CGE model is ideal to study the market-based costs of inaction (or benefits of action) on the economy and the different regions and sectors for the coming decades (until 2060 in this report), the IAM model can be used to study long-term consequences of climate change as well as to explore optimal policy scenarios. Given their level of aggregation, this modelling framework cannot assess the consequences of climate change at the sub-national and local level, even though for some impacts (e.g. extreme events) local consequences far outstrip those at the national and global level. Furthermore, conventional representations of economic agents (households and firms), like in CGE models and IAMs, may not be appropriate if the shock is very large and discontinuous, not only on a local scale, but also in macroeconomic terms.

Large-scale IAMs, most notably those following Nordhaus' DICE and RICE models (Nordhaus and Boyer, 2000; Nordhaus, 2010, 2012; de Bruin et al., 2009a, b; Bosello et al., 2010), have often been used as tools to assess the interactions between economic activity and climate change. These models are constructed to include a stylised representation of as many components as possible of a standard framework to assess the economic costs of climate change. They are based on aggregate damage functions, which are calibrated to the assumed economic baseline and used to subtract the overall costs of climate change from an appropriate economic measure, such as GDP. IAMs are generally based on a forward-looking framework, which can be used to study the trade-off between the ability to adjust emissions and economic growth in anticipation of economic losses due to the impacts of future climate change as well as policy options to reduce climate change damages through adaptation.

However, climate change does not have a uniform effect on different economic activities, and more recently modelling efforts have attempted to reflect this. A key challenge is to adequately capture the heterogeneity of climate change impacts, their geographic occurrence, and response to shifts in climate variables. But, it is also necessary to capture how impacts on natural and human systems vary in character and magnitude across regions and how these translate into shocks to the economy through the channels of different economic variables, with some activities or sectors being more severely affected than others (Sue Wing and Lanzi, 2014).

To address this, a more recent literature has tried to combine economic models that have sectoral details with information obtained from climate models and the empirical literature on climate damages. The models mostly used for this type of assessment are CGE models, which have the characteristic of depicting economic sectors and regions through trade flows and productive activities. Compared to IAMs, these models have a more detailed regional and sectoral structure, which can be used to better link climate impacts to the various economic sectors. However, the inclusion of non-market impacts of climate change is by far not straightforward in a CGE model: they therefore tend to capture a smaller subset of impacts compared to some detailed IAMs – and an even smaller subset of the wider impacts literature. These models also have the disadvantage of being computationally more complicated, as they recalculate an economic equilibrium at each time step. While these models are often dynamic, they are in most cases not based on a forward looking structure. Thus, they do not permit the determination of an optimal level of mitigation. They are also generally used for shorter term analysis. While IAMs can generate projections to the end of the 21st century and beyond, CGE models generally have timeframes up to mid-century for relatively detailed models and out to the end of the century for aggregated models. The shorter timescale is partly due to the fact that it is difficult to obtain reliable information on projected changes in sectoral production and demands and other socioeconomic trends that are needed to calibrate the models.

Both modelling approaches have relative advantages and disadvantages. It is possible to overcome these shortcomings by creating an IAM with sectoral details, but this is computationally complicated and generally means that some of the details of the sectoral and regional characteristics are lost. Alternatively, as done in this report, the two approaches can be combined and used to complement each other. By calibrating the two types of models on the same economic baseline and aligning the climate impacts, they can be used as complementary assessments to study different aspects of the same storylines. Nevertheless, differences between the two models remain. Most importantly, the stylised representation of the economy and damages from climate change in AD-DICE cannot fully replicate the sectoral and regional behaviour underlying the more elaborate approach in ENV-Linkages.

This report combines the analysis of regional and sectoral damages of climate change in the ENV-Linkages model developed by the OECD Environment Directorate (Chateau et al., 2013) with an analysis of long-term consequences and policy action done with the AD-DICE model (de Bruin et al., 2009a, b; Agrawala et al., 2011). The damages from climate change are contrasted with a "no-damage 'baseline' projection", which reflects the trend development of the socioeconomic drivers of economic growth (see Section 2.1); these trends abstract from short-term disruptions and business cycles.

ENV-Linkages is a global dynamic computable general equilibrium (CGE) model that describes how economic activities are linked to each other between sectors and across regions. The version used for the current analysis contains 35 economic sectors (see Table A1.1 in the Annex) and 25 regions (reproduced in Table 1.2), bilateral trade flows and has a sophisticated description of capital accumulation using capital vintages, in which technological advances only trickle down slowly over time to affect existing capital stocks. It also links economic activity to environmental pressure, specifically to GHG emissions. In ENV-Linkages, sectoral and regional economic activities and GHG emissions are projected for the medium- and long-term future, up to 2060, based on socio-economic drivers such as demographic developments, economic growth and development in economic sectors (see Chapter 2).

Table 1.2. **Regions in ENV-Linkages**

Macro regions	ENV-Linkages countries and regions
OECD America	Canada Chile Mexico United States
OECD Europe	EU large 4 (France, Germany, Italy, United Kingdom) Other OECD EU (other OECD EU countries) Other OECD (Iceland, Norway, Switzerland, Turkey, Israel)
OECD Pacific	Australia and New Zealand Japan Korea
Rest of Europe and Asia	China Non-OECD EU (non-OECD EU countries) Russian Federation Caspian region Other Europe (non-OECD, non-EU European countries)
Latin America	Brazil Other Lat. Am. (other Latin-American countries)
Middle East and North Africa	Middle-East North Africa
South and South-East Asia	India Indonesia ASEAN9 (other ASEAN countries) Other Asia (other developing Asian countries)
Sub-Saharan Africa	South Africa Other Africa (other African countries)

Source: ENV-Linkages model.

AD-DICE is based on the well-known integrated assessment model DICE (Nordhaus, 1994, 2012) but it is extended to include an explicit representation of adaptation to climate change. In the model economic production leads to emissions of GHGs but industrial carbon dioxide (CO_2) is the only endogenous gas. Emissions increase the stock of CO_2 in the atmosphere, resulting in climate change, which is represented in the model with changes in atmospheric temperature compared to pre-industrial (1900) levels. The economic consequences of climate change (i.e. climate damages), as measured by the change in GDP, are calculated as a function of temperature changes. Climate damages can be reduced with investments in mitigation, that will reduce CO_2 emissions, or with adjustments to the economy (i.e. adaptation). The model is based on an inter-temporal optimisation framework which can be used to find the optimal balance of capital investments, mitigation investments, adaptation investments, adaptation costs. The model has global coverage. AD-DICE, and its sister model AD-RICE, were also used in previous OECD studies, e.g. Agrawala et al. (2011), to gain insights about longer-term dynamics of climate-economy interactions and the relation between mitigation and adaptation policies. Further details on both models are provided in the Annex.

In order to enhance the comparability of the two models, they have both been calibrated on the same economic baseline, which is briefly outlined in Section 2.1. For OECD countries and the main emerging economies, this based on the OECD long-run aggregate growth scenario to 2060 (OECD, 2014a); for other countries the OECD's ENV-Growth model is used. Beyond 2060, AD-DICE has been calibrated following the growth rates of the business-as-usual scenario of the DICE model. The emission pathways in the two models have also been harmonised in order to increase comparability of results on both climate change impacts and policy results.

Furthermore, the damage function of AD-DICE has been recalibrated to the ENV-Linkages damage projections until 2060. To be precise, the parameters for both climate damages and adaptation have been recalibrated using the sectoral damage information from ENV-Linkages. For longer-term developments, the damage function parameters evolve in line with the original DICE specification, i.e. in the very long run, the damage function replicates the original DICE model.

Notwithstanding the remaining differences between both models, combining them allows this report to present results on different aspects of the economic consequences of climate change. The stylised nature of the AD-DICE model makes it more suitable for explorative scenario analysis. The core of the analysis, however, focuses on the sectoral and regional results obtained with the ENV-Linkages model and derived from a production function approach, which links with as much detail as possible climate impact endpoints with the production function that underlies the structure of the model.

1.3.2. *The production function approach*

A key challenge in modelling the link between climate change impacts and economic activities is to adequately capture the heterogeneity of climate change impacts. These vary in character and magnitude across regions and translate into shocks to the economy with some activities and sectors being more severely affected than others, through the channels of different economic variables.

One way to study this complex system in an economic framework is to link each climate impact to different variables in the production function that represents the activity of a specific industry or group of industries in the basic structure of the model. For a general framework see Sue Wing and Fisher-Vanden (2013) and for an overview of modelling applications see Sue Wing and Lanzi (2014). In a production function, output is produced from distinct inputs (e.g. labour and capital), intermediate commodity inputs and primary resources.

By modelling climate change impacts with a production function approach, it is possible to obtain, as for integrated assessment models, the total economic costs of the selected impacts of climate change on GDP. The overall GDP costs are in turn an indicator of the extent to which climate change has an impact on future economic growth; as in this approach damages can also affect capital stocks, it includes a potential direct effect on the growth rate of the economy. Compared to integrated assessment models in which climate damages are subtracted as a total from GDP, the production function approach can also explain how the composition of GDP is affected over time by climate change: what sectors are most affected (for the impacts that have been assessed) and what changes in production factors mostly contribute to changes in GDP.

Climate impacts have the potential to directly affect sectors' use of labour, capital, intermediate inputs and resources.[1] But they will also affect the productivity of inputs to production. Adverse climate-related shocks to the economy therefore act in the same manner as technological retrogressions, necessitating the use of more inputs to generate a given level of output.

Explicitly linking climate impacts to the sectoral economic variable works well for those impacts that are directly affecting economic markets. For non-market impacts, such a direct link with a part of the production function does not exist, and the damages need to be evaluated separately. In principle, the utility function could be used to incorporate both market and non-market damages in one quantitative framework, but specifying such a utility function is far from obvious and left for future research (see Chapter 3 for more details). Thus, in this report some of the main non-market damages are discussed in a stand-alone fashion in Chapter 3.

Modelling climate impacts with a production function approach relies heavily on the available empirical evidence, but also the opportunities to include this information with the modelling framework. Empirical studies that quantify the effect of climate change impact on the economy are numerous but their comprehensiveness varies in terms of geographical coverage and the impacts they consider (Agrawala and Fankhauser, 2008; OECD 2015). For instance, while there is a very large literature on agricultural damages from climate change, empirical studies on the dependence of energy supply on water availability and how this is affected by climate change are still scarce and limited to a few regions (cf. IEA, 2015).

The availability of empirical evidence also affects the decision on how to model each type of impact (Sue Wing and Lanzi, 2014) in the CGE environment. For example, changes in crop yields due to climate change can be modelled as a uniform shock to all crop sectors or, when more information is available, they can be differentiated between the crop sectors and regions. Similarly, impacts on coastal areas from sea level rise could be modelled in the CGE as a single productivity shock in all sectors or as a reduction in the supply of land together with an increase in non-productive defensive investments in exposed sectors.

Modelling climate damages in CGE models also means that a certain level of market-driven, reactive autonomous adaptation to the damages is inherently modelled. In models with sectoral details and a complex production and trade structure, a change in the productivity of a particular input will trigger substitution responses by producers that alter the use of the various inputs. Substitution is a powerful form of market adaptation once the level of the economy at which impacts manifest themselves is reached. The presence of market adaptation in the model also means that the final estimated costs of climate change impacts can be expected to be lower (or higher) than those estimated if adaptation is not considered (or considered to be optimal), as is often the case in IAMs. This feature also allows modellers to study both the direct effects of climate change and the indirect ones, such as the impacts that take place after trade effects.

The technical difficulty in implementing the detailed analysis of the sectoral and regional climate change feedbacks on economic growth is that the various steps that link the economy to climate change cannot be robustly summarised in a simple damage function, as is often used in IAMs. The economic model is used to create projections of economic growth with sectoral and regional details. The regional and sectoral structure of the models, as well as the energy details, can be exploited to produce projections of GHG emissions so as to obtain an emission pathway. Once the emissions are obtained, a climate module, such as MAGICC (Meinshausen et al., 2011), will translate the emission pathway into emission concentrations and temperature changes. This will then be the input or reference to obtain the needed information on climate damages for that specific scenario.

Figure 1.1. **Linking economic and climate change models**

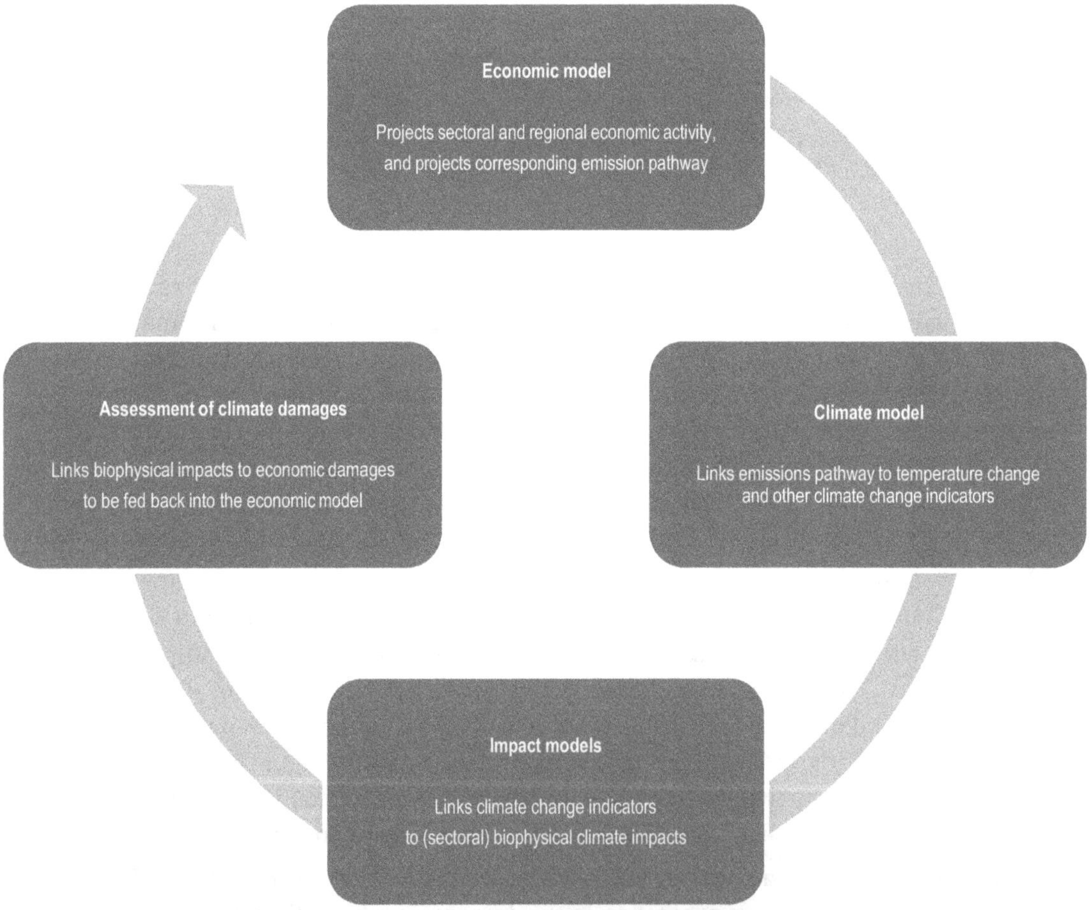

Economic model

Projects sectoral and regional economic activity, and projects corresponding emission pathway

Climate model

Links emissions pathway to temperature change and other climate change indicators

Impact models

Links climate change indicators to (sectoral) biophysical climate impacts

Assessment of climate damages

Links biophysical impacts to economic damages to be fed back into the economic model

Source: Own compilation.

The temperature pathway can be used as an input in two ways. In certain cases it can be used as input for specific sectoral models that will focus on a specific impact, such as a coastal system model for impacts of sea level rise, or an agricultural crop model to obtain crop yield changes. In other cases it can be used as a reference to seek for empirical or modelling studies that have already been done on existing temperature pathways. The existing reference pathways are usually the Representative Concentration Pathways (RCP) (Van Vuuren et al., 2012) or, for older studies, the IPCC Special Report on Emissions Scenarios (SRES) (Nakicenovic and Swart, 2000). In this case the data used will be those relative to the pathway that is closed to the chosen reference scenario.

As a final step in the production function approach, the information obtained on climate damages is fed into the model by sector and region, choosing the most appropriate variables for each climate impact. The final output is a new level of sectoral, regional and global GDP that reflects the costs of climate change on economic growth. Figure 1.1 summarises this process.

1.4. Modelling of sectoral and regional climate impacts

The quantification of climate change impacts in ENV-Linkages relies on available information on how climate impacts affect different economic sectors. The information sources are mostly derived from bottom-up partial-equilibrium models, climate impact models and econometric studies.[2] Table 1.3 provides a summary of the impacts considered and their respective sources from the literature. They refer to the consequences of climate-related changes in agriculture and fisheries, coastal zones, health, and changes in the demand for tourism services and for energy for heating and cooling.

Table 1.3. **Climate impact categories included in ENV-Linkages**

Climate impacts	Impacts modelled	Source	Project	Time frame
Agriculture	Changes in crop yields	IMPACT model – Nelson et al. (2014)	AgMIP	2050
	Changes in fisheries catches	Cheung et al. (2010)	SESAME	2060
Coastal zones	Loss of land and capital from sea level rise	DIVA model – Vafeidis et al. (2008)	ClimateCost	2100
Extreme events	Capital damages from hurricanes	Mendelsohn et al. (2012)		2100
Health	Mortality and morbidity from infectious diseases, cardiovascular and respiratory diseases	Tol (2002)		2060
	Morbidity from heat and cold exposure	Roson and Van der Mensbrugghe (2012) and Ciscar et al. (2014) for Europe	World Bank ENVISAGE model and Peseta II (Europe)	2060
Energy demand	Changes in energy demand for cooling and heating	IEA (2013)	WEO	2050
Tourism demand	Changes in tourism flows and services	HTM – Bigano et al. (2007)	ClimateCost	2100
Ecosystems	No additional impacts covered in the modelling exercise			
Water stress	No additional impacts covered in the modelling exercise			
Tipping points	Not covered in the modelling exercise			

Source: Own compilation.

Most impacts used are assessed for the specific Representative Concentration Pathway (RCP) 8.5 scenario, which describes a pathway of climate change resulting from a fast increase in global emissions. The RCPs were developed by Van Vuuren et al. (2012) and adopted by the IPCC (2013; 2014a, b). Alternatively the impacts are related to the slightly older IPCC A1B SRES scenario (Nakicenovic and Swart, 2000), which describes a future world of very fast economic growth, global population that reaches its maximum number by 2050 and declines thereafter, and the rapid introduction of new and more efficient technologies for all energy sources (IPCC, 2000). The usage of different scenarios introduces only a minor approximation problem in specifying the RCP 8.5 reference, however, because until 2060 the temperature profiles of RCP 8.5 and A1B are reasonably close. Both scenarios are also similar to the ENV-Linkages model baseline with respect to GHG concentrations.

Wherever possible, the central projection uses results from the HadGEM3 model (Madec et al., 1996) from the Hadley Center of the UK Met Office, for the specification of the climate system variables. However, for certain climate impacts the data was only available from other climate models.

All source studies have a global coverage. As most studies come from grid-based data sets and models, they report data with a high spatial resolution, which permits the aggregation of data to match the regional aggregation of the ENV-linkages model. In some cases the source studies specified impact data with a regional aggregation tailored for other CGE models, including the ICES model[3] (Eboli et al., 2010; Bosello et al., 2012; Bosello and Parrado, 2014), which was used as a reference for several climate impacts. The ICES model presents a regional detail very close to that of ENV-Linkages. Simple averaging processes or other simplifying ad hoc assumptions have been used to determine impacts for those few regions not perfectly matching across the two models.

In cases where the data sources were only available until 2050, the trends between 2040 and 2050 have been extrapolated to 2060. In principle, the impacts are not provided for a specific year, but rather for a period of multiple years. Where applicable, the sectoral assessments of impacts for a future period, e.g. a period of 2045-55, have been translated into impacts for the middle year (in this case 2050) and then annual trends have been interpolated for earlier periods when no further information was available.

Two broad categories of climate change impacts can be distinguished. The first affects the supply-side of the economic system, namely the quantity or productivity of primary factors. Land and capital destruction from sea level rise, crop productivity impacts in agriculture, and labour productivity impacts on human health belong to this category. The second category of climate change impacts affects the demand side. Impacts on health expenditures[4] and on energy consumption are of this kind.

1.4.1. *Agriculture*

The climate change impacts on agriculture that are modelled in ENV-Linkages involve sector- and regional-specific changes in crop yields for each of the 8 crop sectors (see Annex I for the sectoral disaggregation of the ENV-Linkages model). The input data on crop yield changes (physical production per hectare) are those shared by the modelling teams involved in the Agricultural Model Intercomparison Project AgMIP (Rosenzweig et al., 2013; Nelson et al., 2014; Von Lampe et al., 2014). This project contains the most robust global assessment of agricultural impacts from climate change published to date. Although impacts on grasslands follow very similar patterns as impacts on crop land, the AgMIP project has not provided information on how grasslands are affected, so impacts on livestock are excluded from the modelling analysis. From the available scenarios shared in the AgMIP project, the central projection uses the HadGEM model, for the specification of the climate system variables, coupled with the DSSAT crop model (Hoogenboom et al., 2012; Jones et al., 2003). The specification of regional climate impacts coming from this model combination was then used as input for the International Food Policy Research Institute's IMPACT model (Rosegrant et al., 2012) to calculate the exogenous yield shocks from changes in crop growth and water stress by water basin. These shocks were then aggregated to the ENV-Linkages model regions. A pathway between 2010 and 2050 was produced by proportionally changing the effect of climate change on the yield growth rate such that in 2050, the yield shocks correspond to the AgMIP projection for the 2050s; this delivers a non-linear impact on yield levels. In line with AgMIP, the estimated yield shocks used in the central projection do not consider the carbon fertilization effects on vegetation as these are deemed too uncertain, although Rosenzweig et al. (2013) do identify it as "a crucial area of research".

To further explore the uncertainty in the assessment of these agricultural impacts, Chapter 2 will also consider the implications of choosing other scenarios for specifying crop yield impacts, based on alternative choices for the underlying crop model, the underlying climate model and – not least – the assumption on CO_2 fertilisation.

Since the IMPACT model does not contain a full production function while ENV-Linkages does, the crop yield shocks have been translated into specific elements in the production functions of the ENV-Linkages agricultural sectors. The yield shocks are implemented in the model as a combination of the productivity of the land resource in agricultural production, and the total factor productivity of the agricultural sectors.[5] This specification mimics the idea that agricultural impacts affect not only purely biophysical crop growth rates but also other factors such as management practices.[6]

Climate change affects crop yields heterogeneously in different world regions. Further, the effects are also not the same for different crops. Figure 1.2 illustrates changes in crop yields at the regional level in 2050 for the central projection using the HadGEM climate model in combination with the DSSAT crop model. This excludes a CO_2 fertilisation effect; the uncertainty related to the choice of climate and crop model, and the effect of CO_2

Figure 1.2. **Impacts of climate change on crop yields in the central projection**

Percentage change in yields in 2050 relative to current climate

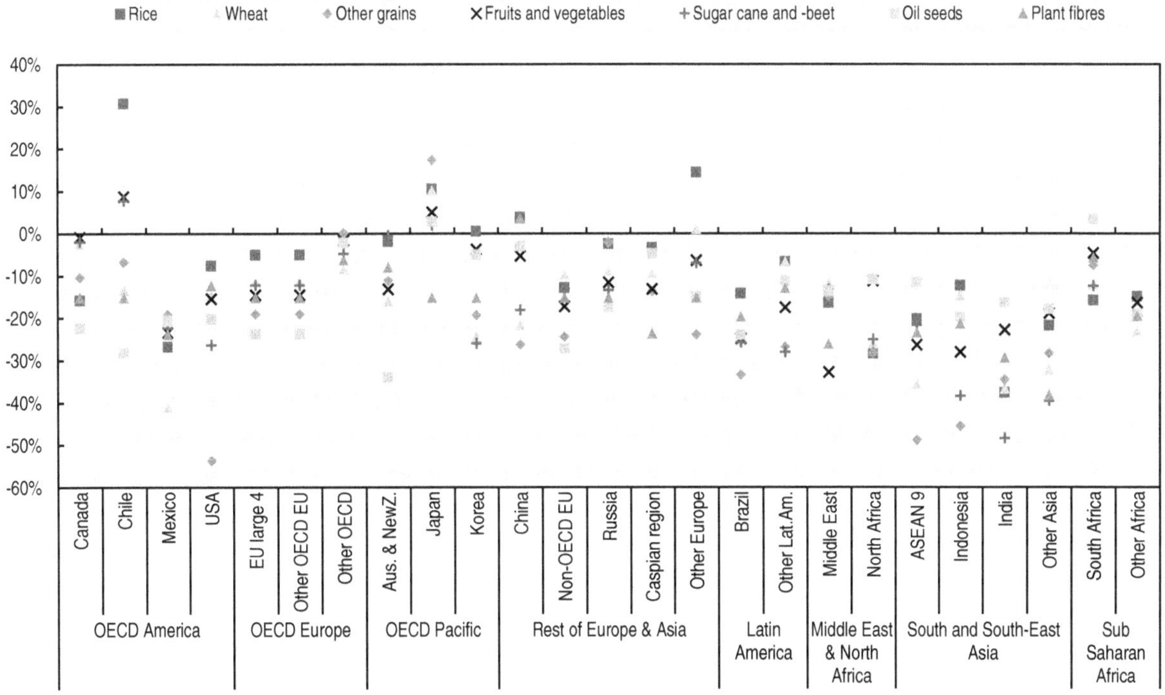

Source: IMPACT model, based on the AgMIP study (Von Lampe et al., 2014).

StatLink http://dx.doi.org/10.1787/888933275901

fertilisation, is further investigated in Chapter 2. While the maps illustrate impacts for 2050, the impacts are not constant over time. They follow a non-linear trend that is extended from 2050 to 2060 using the increase from the previous decade. Impacts for other crops can deviate substantially from the impacts for rice and wheat; they are not reproduced here, but are described in detail in Nelson et al. (2014). Note that these impacts refer to potential shocks: in the CGE model, farmers have options to change their production process and adapt to these shocks and will do so in order to minimise their costs, i.e. market-driven adaptation is endogenously handled inside the economic modelling framework. The modelling framework excludes the possibility to increase the size of irrigated agricultural land. In regions with low water stress levels, this adaptation option can be an important part of the response to climate change (Ignaciuk and Mason-D'Croz, 2014), but is excluded here as markets forces alone are usually insufficient to achieve large-scale expansion of irrigated areas (Ignaciuk, 2015).

Changes in yields of paddy rice by 2050 are strongest in tropical areas, including Central American and Mexico, Saharan African countries, some parts of the Middle East and a large part of South and South-East Asian countries. Some regions have large positive impacts on paddy rice yields. In particular, the highest gains will take place in the Southern parts of Latin America, and particularly in Chile, Japan, and in parts of Easter Europe and continental Asia. Such heterogeneity in impacts suggests that climate change will largely change trade patterns in widely traded commodities such as rice.

Changes in yields of wheat by 2050 are somehow less differentiated, as most regions are negatively affected. The most severe negative impacts take place in Mexico, Western and Eastern Africa, some Southern African countries, Middle East, South and South East Asia, and some Western European regions, such as Belgium, the Netherlands and Germany. While these are the most affected regions, negative impacts are widely spread and also affect most of Europe, continental Asia and North America. Some regions are positively affected by climate change. These include regions with cold climates such as Canada, Russia and Scandinavian countries, most of Central America, Argentina, some countries in Eastern Europe and continental Asia, and a few African countries.

For the *fisheries* sector, the damages reflect projected changes in global fish catch potential caused by climate change. This is modelled in ENV-Linkages as a change in the natural resource stock available to fishing sectors, which approximates the impacts of climate change for fish stocks and the resulting effects for the output of the fisheries sector. Acknowledging that the empirical basis for estimating the impacts on the fisheries sector is very small and uncertainties on projections are very large, the input data used for the modelling is based on one of the most comprehensive assessments, the EU's SESAME project, which in turn uses results from Cheung et al. (2010). This study applies an empirical model (Cheung et al., 2008) that predicts maximum catch potential as dependent upon primary production and distribution. It considers a range of 1066 species of exploited fish and invertebrates. Future projected changes in species distribution are simulated by using a model (Cheung et al., 2008, 2010) that starts with identifying species' preference for environmental conditions and then links them to the expected carrying capacity. The environmental conditions considered include seawater temperature, salinity, distance from sea-ice and habitat types, but the assessment excludes any effects related to ocean acidification. The model assumes that carrying capacity varies positively with habitat suitability of each spatial cell. Finally, the related change in total catch potential is determined aggregating spatially and across species.

The input data for the fisheries sector in the ENV-Linkages model is the percentage change in fish catch with respect to 2000 as described above. The most negatively affected regions by 2060 are North Africa (-27%) and Indonesia (-26%). Some European regions, the Middle East, Chile and several countries in South East Asia have impacts ranging from -10% to -15%. Smaller negative impacts also take place in China, Korea, Brazil and other Latin American countries, Mexico, and some European countries. In some countries fish catches actually increase. The highest increases will occur in Russia (+25%) and in the five major European economies (+23%). Small positive impacts are seen in the United States, Canada, Oceania and the Caspian region. Other world regions (India, other developing countries in Asia, South Africa and the rest of Africa) are basically unaffected.

1.4.2. *Coastal zones*

Coastal land losses due to sea level rise are included in the ENV-Linkages model as changes in the availability of land as well as damages to physical capital. Both modifications concern land and capital stock variables by region in the model. As information on capital losses are not readily available, in line with Bosello et al. (2012), land and capital stock changes are approximated by assuming that changes in capital supply match land losses as a percentage change from baseline.

Estimates of coastal land lost to sea level rise are based on the DIVA model outputs (Vafeidis et al., 2008) as used in the European Union's (EU) FP7 ClimateCost project (Brown et al., 2011) and generated with the HadGEM model. DIVA is a sector model designed to address the vulnerability of coastal areas to sea level rise and other ocean- and river-related events, such as storm surges, changes in river morphology and altered tidal regimes. The model is based on a world database of natural system and socioeconomic factors for world coastal areas reported with spatial details. Changes in natural and socio-economic conditions of possible future scenarios are implemented through a set of impact and adaptation algorithms. Impacts are then assessed both in terms of physical losses (i.e. sq. km of land lost) and economic costs (i.e. value of land lost and adaptation costs).

The regions that are most affected by sea level rise are those in South and South East Asia, with highest impacts in India, and other developing countries in the region. The projected land and capital losses expressed as percentage of total regional agricultural land area in 2060 with respect to the year 2000 are respectively -0.63% for India and -0.86% for the Other Developing Asia region of ENV-Linkages. Other countries in the region are also affected but to a smaller extent. Some impacts are also felt in North America, with Canada, Mexico and the United States being affected. Canada has the highest loss in land (and capital) in this region (-0.47% in 2060 with respect to 2000). Smaller impacts occur in Middle East (-0.35%) and in Europe, where the highest impacts are felt in the aggregate non-OECD Europe region (-0.37%), which includes, among other countries, Israel, Norway and Turkey. Other world regions, such as Africa, South America and continental European regions are on balance hardly affected by sea level rise.

1.4.3. *Extreme events*

There are many types of extreme events and they affect the economy in different ways. However, given the uncertainties involved in the frequency and damages caused by these events and the difficulties in attributing such events to climate change, the available data on how the economy will be affected is still scarce. Recently, the assessment by Mendelsohn et al. (2012) has provided some quantitative assessment and projections on damages from hurricanes that can be used as input in an economic framework. Mendelsohn et al. (2012) stress that the regional damages are quite sensitive to the climate model that is used to project future climate conditions, and projections based on the HadGEM model are not available. Hence, the analysis of the economic consequences was realised in ENV-Linkages on a multi-model average.[7] This may of course dampen some of the more severe consequences projected by individual models.

Mendelsohn et al. (2012) find that climate change is predicted to increase the frequency of high-intensity storms in selected ocean basins as the century progresses, although this depends on the climate model used. These climate-induced damages are included in ENV-Linkages as reductions in regional capital stocks from tropical cyclones. Due to lack of information, this is assumed to affect all sectors (which is in line with the normal CGE model assumption that new investments in capital are fully malleable across the economy).

Mendelsohn et al. (2012) also find that the current annual global damage from tropical cyclones is USD 26 billion, which is equivalent to 0.04% of global GDP, and roughly double (in absolute amount) by the end of the century under current climate conditions, i.e. due to changes in socioeconomic conditions.[8] However, these damages are projected to double again by the end of the century due to climate change. Most additional climate-induced damages are predicted to take place in North America, East Asia and the Caribbean-Central American region, where the United States, Japan and China will be most affected.

1.4.4. *Health*

Within the health impact category, the ENV-Linkages model covers both climate-related illnesses and effects related to heat stress. Impacts on human health linked with *climate-related diseases* are expressed by changes in mortality and morbidity, following Bosello et al. (2012) and Bosello and Parrado (2014).[9] The illnesses considered are vector-borne diseases (malaria, schistosomiasis and dengue), diarrhoea, cardiovascular and respiratory diseases. The assessment for cardiovascular diseases includes both cold and heat stress. Within the production function approach, the modelling technique for these health impacts is to translate the results of the empirical literature into changes in labour productivity and demand for health services (Bosello et al., 2006) – explicitly excluding the welfare (or "disutility") impacts of premature deaths from climate change. While there are other factors that are affected by these illnesses, labour productivity is the most suitable variable to capture the effects that climate-related diseases have on the economy.

Estimates of the change in mortality due to vector-borne diseases are taken from Tol (2002), which are based on modelling studies (Martens et al., 1995, 1997; Martin and Lefebvre, 1995; Morita et al., 1994) as well as on mortality and morbidity figures from the World Health Organization's Global Burden of Disease data (Murray and Lopez, 1996).[10] These studies suggest that the relationship between climate change and malaria is linear. This relationship is also applied to schistosomiasis and dengue fever. To account for changes in vulnerability possibly induced by improvement in living standards, Tol (2002) applies a relationship between per capita income and disease incidence (Tol and Dowlatabadi, 2001). This relationship is used to assess the impacts for the CIRCLE baseline by using the projected per capita regional income growth of the ENV-Linkages model (see Chapter 2).

For diarrhoea, an estimated equation describes how increased temperatures increase both mortality and morbidity, while negative income elasticities imply lower impacts with rising income, with mortality declining more rapidly than morbidity (Link and Tol, 2004). For premature deaths due to cardiovascular and respiratory diseases, data are based on a meta-analysis performed in 17 countries (Martens, 1998). Tol (2002) extrapolates these findings to all other countries, using the current climate as the main predictor. Cardiovascular (for both cold and heat stress) and respiratory mortality (for heat stress only) are assumed to only affect urban population.

The resulting changes in labour productivity from climate-induced diseases, which have been summarised in Bosello et al. (2012), are used as in input in ENV-Linkages. By 2060, the highest negative effects take place in Africa and the Middle East (-0.6% for South Africa, -0.5% for North Africa and the Middle East and -0.4% for other African countries). Smaller impacts take place in Brazil, Mexico and in developing countries in Asia (-0.3%), as well as in Indonesia, the United States, South-East Asia and most of Latin America (-0.2%). Some regions are projected to have positive impacts on labour productivity from climate-induced diseases, the highest being in Russia (+0.5%), Canada (+0.4%) and China (+0.2%). In other regions the impacts are either very small or inexistent.

Changes in *health care expenditures* for climate-related diseases are also taken from Bosello et al. (2012). The costs of vector borne diseases are based on Chima et al. (2003), who report the expenditure on prevention and treatment costs per person per month. Changes in health expenditure are small as percentage of GDP.[11] In 2060, they are projected to be highest in the developing countries in Asia (0.5%), in Brazil and in the Middle East and North Africa region (0.3%). Additional demands for health services are very small in other regions. Interestingly, they are negative in Canada and in large EU economies, such as Germany and France (-0.1%), where reduced cardiovascular disease expenditures dominate.

Occupational heat stress is modelled as having an impact on labour productivity. This builds on Kjellstrom et al. (2009), who identify a link between the global temperature, heat and humidity, and work ability for different types of activities (agriculture, industry and services). Ideally one would combine the sectoral reductions in work ability to the regional temperature increases to identify labour productivity losses. Unfortunately, there is insufficient data to do so. However, data derived from Roson and Van der Mensbrugghe (2012) translates the underlying regional climate profiles into labour productivity losses as a function of global average temperature increase. For the European regions, data from Ciscar et al. (2014) are adopted.

Until now, most assessments (Eboli et al., 2010, Ciscar et al., 2014) have first aggregated the various productivity losses across sectors and then apply these averages to all economic sectors. Recent research (e.g. Graff Zivin and Neidell, 2014; Somanathan et al., 2014) has highlighted that productivity of sectors where workers are mostly outdoor (i.e. heat-exposed industries) are on balance much more affected by increased heat. For indoor activities, in turn, the most severe consequences can be avoided by increased air conditioning, which is at least partially captured under the impacts of changed energy demand (see e.g. Somanathan et al., 2014). Hence, the impacts are assumed to be limited to heat-exposed sectors. In line with Ciscar et al. (2014), labour productivity losses are concentrated in the agricultural, forestry, fisheries, and construction sectors, and exclude most manufacturing and services sectors (see Annex I for a full list of sectors in ENV-Linkages).

The highest impacts on labour productivity caused by heat stress in 2060 take place in regions with relative large proportions of outdoor workers and warm climates. The most severely affected regions, with productivity losses between 3 and 5%for outdoor activities for a one degree temperature increase are in non-OECD, non-EU European countries, Latin America (incl. Brazil and Chile), Mexico, China, Other Developing Asia and South Africa. Most OECD countries, including USA, Japan and the OECD EU countries have much smaller effects, of less than 1%.[12]

The health impacts of climate change have economic consequences that go beyond market costs. These costs, such as the costs of premature deaths, cannot be accounted for in the ENV-Linkages model. However, they can be evaluated using WTP techniques and, for premature deaths, the Value of a Statistical Life. These impacts are further discussed in Chapter 3.

1.4.5. *Energy demand*

Residential *energy demand* has been projected to change due to climate change. As discussed in Chapter 4, the energy sector is heavily influenced by mitigation policies. But even in the absence of mitigation policies will there be impacts on energy demand and supply. Changes in households' *demand for oil, gas, coal and electricity* from less energy consumption for heating and more for cooling, have been captured directly in the model as a change in consumer demand for the output of these energy services. Changing residential energy demand in response to climate change is derived from the IEA, which provides data on space heating and cooling by carrier until 2050 under its Current Policies Scenario (IEA, 2013a). Data until 2060 was extrapolated using trends in demand from 2040 to 2050. The IEA derives its projections from its World Energy Model, which is a large-scale partial-equilibrium model designed to replicate the functioning of energy markets over the medium- to long-term. It determines future energy supply and demand for different energy carriers (supply: coal, oil, natural gas, biomass; demand: coal, oil, natural gas, nuclear, hydro, bioenergy, and other renewables) according to trends in energy prices, CO_2 prices, technologies, and socioeconomic drivers. The baseline trends without climate change are characterised by increased heating and cooling demand for most economies, driven to a large extent by higher incomes; the baseline also projects a strong trend of electrification, which affects especially heating energy demand. Demand for space heating and cooling under climate change is affected by factors including the anticipated change in heating and cooling degree days due to climate change (IEA, 2013; IEA, 2014).

Overall, global energy demand for space cooling is projected to grow by roughly 250% between 2010 and 2060 under no climate change and by 330% if climate change is taken into account. Increases in demand for heating are much lower, with a projected 42% rise until 2060 without climate change and a 16%-increase in the climate change scenario, but start from a much larger base level. Non-OECD countries drive most of the increase in demand both for heating and cooling, and particularly so in heating. By 2060, household demand for cooling is projected to be 27% of the total demand for space heating and cooling purposes as compared to 9% in 2010.

Climate change-induced shifts in the demand for electricity until 2060 can reflect both an increased demand for cooling purposes, i.e. air conditioning, as well as decreased demand for electric space heating as a response to higher average temperatures. Globally, total annual electricity demand is projected to remain largely unaffected by climate change by 2060, with increases in demand for cooling during summers balancing decreases in demand for heating during winters. Of the 25 regions in ENV-Linkages, about half are projected to increase their total demand for electricity due to greater need for cooling, including the EU7 (+11.4%), Chile (+8.4%) and non-OECD EU countries (+7.3%). Korea (-6.7%), non-EU OECD Europe (-3.7%), and the Caspian region (-3.1%) are part of the other half of the regions in which decreased consumption of electricity for heating will more than offset increases in demand for cooling (given the strong trend towards electrification of heating in the baseline without climate change). In the vast majority of the regions, positive or negative variations in demand for cooling and heating under climate change are projected to stay below 3% relative to the baseline total electricity demand of households.

As global temperatures increase with climate change, the IEA projections suggest that climate change will lead to reduced household demand for all major space heating fuels, i.e. gas, oil and coal, relative to the baseline. In total household demand for gas, change in gas-based space heating is projected to fall by about 7%; oil-based space heating by 1% (out of total household demand for oil); and coal-fired heating by 17% (out of total household demand for coal). If aggregated, the demand for gas, oil, and coal for heating purposes is projected to decline by approximately 3% of total household demand for these fuels (or decline by 2% out of total energy demand for heating and cooling if electricity is included). At the regional level, changes can be more pronounced: the demand of gas, oil and coal in total household demand for these fuels is going to decrease most by 2060 in other EU countries (-19.8%), OCE (-17.6%), EU7 (-13.5%), and Chile (-12.5%). The other regions will experience falling demand for these fuels by less than 10% of household fuel demand (with 12 regions reducing demand by less than 2.5%).

While the consequences of climate change on space cooling and heating demand are captured in the model, the impacts of climate change on *energy supply* (such as interruptions in the availability of cooling water for thermal electricity generation due to heat extremes or droughts, the change in water availability for hydro-electricity, etc.) are not included in the analysis. The direct impacts of climate change on energy supply are not included in the analysis, but the CGE model does capture the endogenous effects on energy markets induced by the impacts on demand, and cause changes in energy prices and supply. As with other demand impacts, any endogenous shifts in demand between carriers, as a result of changing prices and income levels, are also fully captured in the model.

1.4.6. *Tourism demand*

Changes in *tourism* flows reflect projected changes in tourist destinations due to changes in climate. For instance, projected decreases in snow cover in the Alps in the future might lead tourists to go skiing in other regions. These changes in the regional demand for tourism services are derived from simulations based on the Hamburg Tourism Model (Bigano et al., 2007); it thus contains only one projection and alternative models may provide different projections, especially at the regional scale. However, this approach has been amply used in EU research projects and in previous applications in CGE models (Berrittella et al., 2006 and Bigano et al., 2008). The Hamburg Tourism Model is an econometric simulation model that projects domestic and international tourism by country. The share of domestic tourists in total tourism depends on the climate in the home country and on per capita income. Climate change is represented in this simple approximation by annual mean temperature. A number of other variables, such as country size, are included in the estimation, but these factors are held constant in the simulation. International tourists are allocated to all other countries on the basis of a general attractiveness index, climate, per capita income in the destination countries, and the distance between origin and destination. Total tourism expenditure is then calculated multiplying the number of tourists times an estimated value of the average individual expenditure.

In the ENV-Linkages model, climate damages from tourism have been modelled by modifying the quality of tourism services in different regions. In contrast to Bosello et al. (2012), changes in demand for tourism are not forced changes in household expenditures, but they are rather induced by a change in the quality of the service provided. Such quality changes are represented in the model as a change in total factor productivity of the tourism services sector. Changes in tourism expenditures are largely a shift between countries, plus a shift from tourism to other commodities in domestic consumption, as changes in domestic tourism flows do not affect economy-wide expenditures. On balance, global expenditures on tourism are decreasing, implying a net negative impact on the global economy. The main reason for the reduction of global tourism expenditures is that the average quality of tourism services goes down, and to some extent consumers respond to the increase price of tourism to shift towards other consumption categories.

The regional effects are crucial for tourism, as some countries are negatively and others positively affected. The countries with the highest gains in tourism expenditure by 2060, expressed as percentage change with respect to the baseline scenario, are Canada (+92%), Russia (+66%) and the United States (+21%). Smaller positive impacts (of around +10%) also occur in Chile and Japan. The largest negative impacts instead take place in Latin America, excluding Chile and Brazil (-27%), Mexico (-25%), as well as Africa, excluding South Africa, China, South-East Asia and developing countries in Asia (with impacts around -20%). Smaller negative impacts take place in South Africa (-14%), Indonesia (-13%) the EU OECD regions (-9%) and the Caspian region (-7%). Other countries have much smaller impacts.

Notes

1. An example is loss of coastal land, buildings and infrastructure due to inundation as a result of sea level rise.
2. Much of the information used is an elaboration of data provided by recently concluded and ongoing research projects, including both EU Sixth and Seventh Framework Programs (FP6 and FP7) such as ClimateCost, SESAME and Global-IQ and model inter-comparison exercises such as AgMIP. These data have been kindly provided by the researchers involved in these projects.
3. The ICES model is operated by the Euro-Mediterranean Centre for Climate Change (CMCC), Italy. For detailed information about the model please refer to the ICES website: *www.cmcc.it/models/ices-intertemporal-computable-equilibrium-system.*
4. Health impacts are calculated with a cost of inaction approach, which does not account for other costs to society. A valuation of full economic impacts would imply higher costs.
5. Due to a lack of further information, the percentage change in yields is attributed equally to both parts.
6. Note that labour productivity changes in agriculture due to heat stress are captured in the health category.
7. The emission projection used by Mendelsohn et al. (2012) is the SRES A1B scenario, which leads to somewhat lower projected climate change than the CIRLE baseline; this difference implies that the hurricane damages presented here are slightly underestimated but this effect is ignored as the differences until 2060 are small.
8. Note that Hsiang and Jina (2014) find much larger impacts of tropical cyclones on the future global economy by focusing on the consequences for long-term economic growth.

9. Mortality effects have not been incorporated in the model for other climate damages. However, in the case of diseases, it was not possible to disentangle morbidity and mortality effects, so they have been included in the assessment.

10. It is acknowledged that this implies that the assessment cannot take recent developments in the literature, such as updates to the Global Burden of Disease study, into account. Unfortunately, a full updated assessment of the effects of climate change on disease-related health impacts is beyond the scope of the current study.

11. Using explained in Section 1.3, additional health care costs are not directly subtracted from GDP, but rather represented as a forced expenditure by households and the government. Indirectly, this affects GDP.

12. In some cases, there are large differences within the aggregated regions. For example, on the OECD EU countries, the productivity losses are largely concentrated in the Mediterranean countries.

References

Agrawala, S. and S. Fankhauser (2008), "Putting Climate Change Adaptation in an Economic Context", in *Economic Aspects of Adaptation to Climate Change: Costs, Benefits and Policy Instruments*, OECD Publishing, Paris, *http://dx.doi.org/10.1787/9789264046214-3-en.*

Agrawala, S. et al. (2011), "Plan or React? Analysis of adaptation costs and benefits using Integrated Assessment Models", *Climate Change Economics*, Vol. 2(3), pp. 175-208.

Berrittella, M. et al. (2006), "A general equilibrium analysis of climate change impacts on tourism", *Tourism Management*, Vol. 25, pp. 913-924.

Bigano, A., J.M. Hamilton and R.S.J. Tol (2007), "The Impact of Climate Change on Domestic and International Tourism: A Simulation Study", *Integrated Assessment Journal*, Vol. 7, pp. 25-49.

Bigano, A. et al. (2008), "Economy-wide impacts of climate change: a joint analysis for sea level rise and tourism", *Mitigation and Adaptation Strategies for Global Change*, Vol. 13(8), pp. 765-791.

Risky Business Project (2014), *The economic risks of climate change in the United States: A climate risk assessment for the United States*, *www.riskybusiness.org*.

Bosello, F. and R. Parrado (2014), "Climate Change Impacts and Market Driven Adaptation: the Costs of Inaction Including Market Rigidities", *FEEM Working Paper*, No. 64.2014.

Bosello, F., F. Eboli and R. Pierfederici (2012), "Assessing the Economic Impacts of Climate Change. An Updated CGE Point of View", *FEEM Working Paper*, No. 2.2012.

Bosello, F., C. Carraro and E. De Cian (2010), "Climate Policy and the Optimal Balance between Mitigation, Adaptation and Unavoided Damage", *Climate Change Economics*, 1: pp. 71-92.

Bosello, F., R. Roson and R.S.J. Tol (2006), "Economy wide estimates of the implications of climate change: human health", *Ecological Economics*, Vol. 58, pp. 579-591.

Braconier, H., G. Nicoletti and B. Westmore (2014), "Policy Challenges for the Next 50 Years", *OECD Economic Policy Papers*, No. 9, OECD Publishing, Paris, *http://dx.doi.org/10.1787/5jz18gs5fckf-en*.

Brown S. et al. (2011), "The Impacts and Economic Costs of Sea-Level Rise in Europe and the Costs and Benefits of Adaptation. Summary of Results from the EC RTD ClimateCost Project", in: P. Watkiss, (ed.), *The ClimateCost Project. Final Report. Volume 1: Europe*, Stockholm Environment Institute, Sweden.

Chateau, J., R. Dellink and E. Lanzi (2014), "An Overview of the OECD ENV-Linkages Model: Version 3", *OECD Environment Working Papers*, No. 65, OECD Publishing, Paris, *http://dx.doi.org/10.1787/5jz2qck2b2vd-en*.

Cheung, W.W.L., V.W.Y. Lam, J.L. Sarmiento, K. Kearney, R. Watson, R., D. Zeller and D. Pauly (2010), "Large-scale redistribution of maximum fisheries catch potential in the global ocean under climate change", *Global Change Biology*, Vol. 16, pp. 24-35.

Cheung, W.W.L., V.W.Y. Lam and D. Pauly (2008), "Dynamic bioclimate envelope model to predict climate-induced changes in distribution of marine fishes and invertebrates", in: *Modelling Present and Climate-Shifted Distributions of Marine Fishes and Invertebrates*, Fisheries Centre Research Reports 16(3) (W.W.L. Cheung, V.W.Y. Lam and D. Pauly, eds.), pp. 5-50. University of British Columbia, Vancouver.

Chima, R.I., C.A. Goodman and A. Mills (2003), "The economic impact of malaria in Africa: A critical review of the evidence", *Health Policy*, Vol. 63, pp. 17-36.

Ciscar, J.C. et al. (2011), "Physical and economic consequences of climate change in Europe", *Proceedings of the National Academy of Science (PNAS)*, Vol. 108, pp. 2678-83.

Ciscar. J.C. et al. (2014), "Climate impacts in Europe. The JRC PESETA II project", *JRC Scientific and Policy Reports*, No. EUR 26586EN, Luxembourg: Publications Office of the European Union.

De Bruin, K.C., R.B. Dellink and R.S.J. Tol (2009a), "AD-DICE: An implementation of adaptation in the DICE model", *Climatic Change*, Vol. 95, pp. 63-81.

De Bruin, K., R. Dellink and S. Agrawala (2009b), "Economic Aspects of Adaptation to Climate Change: Integrated Assessment Modelling of Adaptation Costs and Benefits", *OECD Environment Working Papers*, No. 6, OECD Publishing, Paris, *http://dx.doi.org/10.1787/225282538105*.

Dell, M., B.F. Jones and B.A. Olken (2009), "Temperature and income: Reconciling new cross-sectional and panel estimates", *American Economic Review*, Vol. 99(2), pp. 198-204.

Dell, M., B.F. Jones and B.A. Olken (2013), "What do we learn from the weather? The new climate-economy literature", *NBER (National Bureau of Economic Research) Working Paper Series*, No. 19578, NBER, Cambridge, Massachusetts.

Dellink, R. et al. (2014), "Consequences of Climate Change Damages for Economic Growth: A Dynamic Quantitative Assessment", *OECD Economics Department Working Papers*, No. 1135, OECD Publishing, Paris, *http://dx.doi.org/10.1787/5jz2bxb8kmf3-en*.

Eboli, F., R. Parrado and R. Roson (2010), "Climate-change feedback on economic growth: Explorations with a dynamic general equilibrium model", *Environment and Development Economics*, Vol. 15, pp. 515-533.

EPA (2015), "Ecosystems: Climate impacts on ecosystems", website article, United States Environmental Protection Agency, *www.epa.gov/climatechange/impacts-adaptation/ecosystems.html* (accessed 24 March 2015).

Hsiang, S.M. and A.S. Jina (2014), "The causal effect of environmental catastrophe on long-run economic growth", *NBER Workshop Paper*, No. 20352.

Hoogenboom, G. et al. (2012), "Decision support system for agrotechnology transfer (DSSAT) version 4.5", CD-ROM. University of Hawaii, Honolulu, Hawaii.

Garnaut, R. (2008), *The Garnaut Climate Change Review: Final Report*, Cambridge University Press.

Garnaut, R. (2011), *The Garnaut review 2011: Australia in the global response to climate change*, Cambridge University Press.

Graff Zivin, J. and M. Neidell (2014), "Temperature and the Allocation of Time: Implications for Climate Change", *Journal of Labor Economics*, 32: 1-26.

IEA (2015), *World Energy Outlook Special Report on Energy and Climate Change*, IEA, Paris.

IEA (2014), *World Energy Outlook*, IEA, Paris, *http://dx.doi.org/10.1787/weo-2014-en*.

IEA (2013), *Redrawing the energy climate map*, IEA, Paris.

Ignaciuk, A. (2015), "Adapting Agriculture to Climate Change: A Role for Public Policies", *OECD Food, Agriculture and Fisheries Papers*, No. 85, OECD Publishing, Paris, *http://dx.doi.org/10.1787/5js08hwvfnr4-en*.

Ignaciuk, A. and D. Mason-D'Croz (2014), "Modelling Adaptation to Climate Change in Agriculture", *OECD Food, Agriculture and Fisheries Papers*, No. 70, OECD Publishing, Paris, *http://dx.doi.org/10.1787/5jxrclljnbxq-en*.

IPCC (2014a), *Climate Change 2014: Impacts, Adaptation, and Vulnerability. Part A: Global and Sectoral Aspects*, Contribution of Working Group II to the Fifth Assessment Report of the Intergovernmental Panel on Climate Change [Field, C.B., V.R. Barros, D.J. Dokken, K.J. Mach, M.D. Mastrandrea, T.E. Bilir, M. Chatterjee, K.L. Ebi, Y.O. Estrada, R.C. Genova, B. Girma, E.S. Kissel, A.N. Levy, S. MacCracken, P.R. Mastrandrea, and L.L.White (eds.)]. Cambridge University Press, Cambridge, United Kingdom and New York, NY, USA, 1132 pp.

IPCC (2014b), *Climate Change 2014: Mitigation of Climate Change*, Contribution of Working Group III to the Fifth Assessment Report of the Intergovernmental Panel on Climate Change [Edenhofer, O., R. Pichs-Madruga, Y. Sokona, E. Farahani, S. Kadner, K. Seyboth, A. Adler, I. Baum, S. Brunner, P. Eickemeier, B. Kriemann, J. Savolainen, S. Schlömer, C. von Stechow, T. Zwickel and J.C. Minx (eds.)]. Cambridge University Press, Cambridge, United Kingdom and New York, NY, USA.

IPCC (2013), *Climate Change 2013: The Physical Science Basis*, Contribution of Working Group I to the Fifth Assessment Report of the Intergovernmental Panel on Climate Change [Stocker, T.F., D. Qin, G.-K. Plattner, M. Tignor, S.K. Allen, J. Boschung, A. Nauels, Y. Xia, V. Bex and P.M. Midgley (eds.)]. Cambridge University Press, Cambridge, United Kingdom and New York, NY, USA, 1535 pp.

Jones, J.W. et al. (2003), DSSAT Cropping System Model, *European Journal of Agronomy*, Vol. 18, pp. 235-265.

Kjellstrom T. et al. (2009), "The direct impact of climate change on regional labor productivity", *Archives of Environmental Occupation and Health*, Winter; 64(4), pp. 217-27.

Link, P.M. and R.S.J. Tol (2004), "Possible economic impacts of a shutdown of the thermohaline circulation: An application of *FUND*", *Portuguese Economic Journal*, Vol. 3, pp. 99-114.

Madec, G. and M. Imbard (1996), "A global ocean mesh to overcome the north pole singularity", *Climate Dynamics*, Vol. 12, p381-388.

Martens, W.J.M., T.H. Jetten, J. Rotmans and L.W. Niessen (1995), "Climate change and vector-borne diseases: A global modelling perspective", *Global Environmental Change*, Vol. 5(3), pp. 195-209.

Martens, W.J.M., T.H. Jetten and D.A. Focks (1997), "Sensitivity of malaria, schistosomiasis and dengue to global warming", *Climatic Change*, Vol. 35, pp. 145-156.

Martens, W.J.M. (1998), "Health impacts of climate change and ozone depletion: An ecoepidemiologic modelling approach", *Environmental Health Perspectives*, Vol. 106(1), pp. 241-251.

Martin, P.H. and M.G. Lefebvre (1995), "Malaria and Climate: Sensitivity of Malaria Potential Transmission to Climate", *Ambio*, Vol. 24(4), pp. 200-207.

Meinshausen, M., S.C.B. Raper and T.M.L. Wigley (2011), "Emulating coupled atmosphere-ocean and carbon cycle models with a simpler model, MAGICC6: Part I – Model Description and Calibration", *Atmospheric Chemistry and Physics*, Vol. 11, pp. 1417-1456.

Mendelsohn, R., K. Emanuel, S. Chonabayashi, and L. Bakkensen (2012), "The Impact of Climate Change on Global Tropical Cyclone Damage", *Nature Climate Change* 2: 205-209.

Morita, T., M. Kainuma, H. Harasawa, K. Kai and Y. Matsuoka (1994), "An Estimation of Climatic Change Effects on Malaria", *Working Paper, National Institute for Environmental Studies*, Tsukuba.

Murray, C.J.L. and A.D. Lopez (1996), *Global Health Statistics*, Harvard School of Public Health, Cambridge.

Nakicenovic, N. and R. Swart (eds.) (2000), *Special Report on Emissions Scenarios. A Special Report of Working Group III of the Intergovernmental Panel on Climate Change*. Cambridge University Press: Cambridge, UK and New York.

Nelson, G.C. et al. (2014), "Climate change effects on agriculture: Economic responses to biophysical shocks", *Proceedings of the National Academy of Sciences*, Vol. 111(9), pp. 3274-3279.

Nordhaus, W.D. (2012), "Integrated Economic and Climate Modeling,", in: P. Dixon and D. Jorgenson (eds.), *Handbook of Computable General Equilibrium Modeling*, Elsevier.

Nordhaus, W.D. (2010), "Economic aspects of global warming in a post-Copenhagen environment", *Proceedings of the National Academy of Sciences*, Vol. 107(26), pp. 11721-11726.

Nordhaus, W.D. (2007), *A question of balance*, Yale University Press, New Haven, United States.

Nordhaus, W.D. (1994), *Managing the Global Commons: The Economics of the Greenhouse Effect*, MIT Press, Cambridge, MA.

Nordhaus W.D. and J. Boyer (2000), *Warming the World*. Economic Models of Global Warming, The MIT Press, Cambridge, MA.

OECD (2015), *Climate Change Risks and Adaptation: Linking Policy and Economics*, OECD Publishing, Paris, *http://dx.doi.org/10.1787/9789264234611-en*.

OECD (2014a), *OECD Economic Outlook*, Vol. 2014/1, OECD Publishing, Paris, *http://dx.doi.org/10.1787/eco_outlook-v2014-1-en*.

OECD (2014b), *Climate Change, Water and Agriculture: Towards Resilient Systems, OECD Studies on Water*, OECD Publishing, Paris, *http://dx.doi.org/10.1787/9789264209138-en*.

OECD (2012), *OECD Environmental Outlook to 2050: The Consequences of Inaction*, OECD Publishing, Paris, *http://dx.doi.org/10.1787/9789264122246-en*.

Parson, E.A. and K. Fisher-Vanden (1997), "Integrated Assessment Models of Global Climate Change", *Annual Review of Energy and Environment*, 1997, 22, 589-628.

Rosegrant, M.W. and the IMPACT Development Team (2012), "International model for policy analysis of agricultural commodities and trade (IMPACT): Model Description", International Food Policy Research Institute (IFPRI), Washington, DC (2), *www.ifpri.org/sites/default/files/publications/impactwater2012.pdf.*

Rosenzweig, C. et al. (2013), "Assessing agricultural risks of climate change in the 21st century in a global gridded crop model intercomparison", *Proceedings of the National Academy of Sciences (PNAS)*, Vol. 111(9), pp. 3268-73.

Roson, R. and D. van der Mensbrugghe (2012), "Climate change and economic growth: Impacts and interactions", *International Journal of Sustainable Economy*, Vol. 4, pp. 270-285.

Somanathan, E., R. Smanathan, A. Sudarshan and M. Tewari (2014), "The impacts of temperature on productivity and labor supply: Evidence from Indian manufacturing", *Indian Statistical Institute Discussion Paper* 14-10, Delhi.

Steininger, K.W. et al. (2015), *Economic Evaluation of Climate Change Impacts: Development of a Cross-Sectoral Framework and Results for Austria*, Springer International Publishing, Switzerland.

Stern, N. (2007), *The Economics of Climate Change: The Stern Review*, Cambridge and New York: Cambridge University Press.

Sue Wing, I. and K. Fisher-Vanden (2013), "Confronting the Challenge of Integrated Assessment of Climate Adaptation: A Conceptual Framework", *Climatic Change*, Vol. 117(3), pp. 497-514.

Sue Wing, I. and E. Lanzi (2014), "Integrated assessment of climate change impacts: Conceptual frameworks, modelling approaches and research needs", *OECD Environment Working Papers*, No. 66, OECD Publishing, Paris, *http://dx.doi.org/10.1787/5jz2qcjsrvzx-en.*

Tol, R.S.J. and H. Dowlatabadi (2001), "Vector-borne diseases, climate change, and economic growth", *Integrated Assessment*, Vol. 2, pp. 173-181.

Tol, R.S.J. (2002), "New Estimates of the Damage Costs of Climate Change, Part I: Benchmark Estimates", *Environmental and Resource Economics*, Vol. 21 (1), pp. 47-73.

Tol, R.S.J. (2005), "Emission abatement versus development as strategies to reduce vulnerability to climate change: an application of FUND", *Environment and Development Economics*, Vol. 10(5), pp. 615-629.

US Interagency Working Group on Social Cost of Carbon (2010), "Social cost of carbon for regulatory impact analysis – under executive order 12866", *Technical Support Document.*

US Interagency Working Group on Social Cost of Carbon (2013), "Technical update of the social cost of carbon for regulatory impact analysis – under executive order 12866", *Technical Support Document.*

Vafeidis, A.T., R.J. Nicholls, L. McFadden, R.S.J. Tol, J. Hinkel, T. Spencer, P.S. Grashoff, G. Boot and R.J.T. Klein (2008), "A new global coastal database for impact and vulnerability analysis to sea level rise", *Journal of Coastal Research*, Vol. 24(4), pp. 917-924.

Van Vuuren, D.P., K. Riahi, R. Moss, J. Edmonds, A. Thomson, N. Nakicenovic, T. Kram, F. Berkhout, R. Swart, A. Janetos, S.K. Rose and N. Arnell (2012), "A proposal for a new scenario framework to support research and assessment in different climate research communities", *Global Environmental Change*, Vol. 22(1), pp. 21-35.

Von Lampe, M.D. et al. (2014), "Why do global long-term scenarios for agriculture differ? An overview of the AgMIP Global Economic Model Intercomparison", *Agricultural Economics*, Vol. 45(1), pp. 3-20.

WHO (2012), *Atlas of Health and Climate*, World Health Organization and World Meteorological Organization, Geneva.

WHO (2014), *Quantitative Risk Assessment of the Effects of Climate Change on Selected causes of Death, 2030s and 2050s*, World Health Organization, Geneva.

Chapter 2

The damages from selected climate change impacts to 2060

This chapter first briefly outlines the main socioeconomic trends that are projected to emerge regardless of climate change or climate change policies. It then presents the results of the numerical evaluation of the economic costs of climate change until 2060 using the ENV-Linkages model. The focus of this assessment of the damages from climate change is on market impacts and macroeconomic consequences, but the chapter also investigates consequences for specific regions, the sectoral structure of the different economies, and the consequences for international trade.

Using the framework presented in Chapter 1, the ENV-Linkages model can be used to bring together the assessment of selected climate change impacts and impacts on different drivers of economic growth into one consistent analytical framework. This framework is applied to assess the projected consequences of those climate change impacts on the different elements in the economy, not least Gross Domestic Product (GDP). Based on the evaluation of the impacts that can be accounted for in the modelling framework, projections until 2060 can be made of how the regional economies may evolve over time under current ("no-damage") climate conditions and when climate change damages are factored in.

2.1. The no-damage "baseline" projection

The analysis of projected long-term costs of climate change is built on a baseline projection. Such a baseline projection is characterised by an absence of new climate policies, the continuation of current policies for other policy domains (including energy) and plausible socio-economic developments, including demographic trends, urbanisation and globalisation trends.[1] A baseline projection is not a prediction of what will happen, but rather a plausible scenario describing a certain storyline for how these key trends affect future economic development in the absence of unexpected shocks. A typical "business-as-usual" baseline projection should include the damages from climate change, because they will occur regardless of policy action and affect the economy anyway. However, to assess the costs of inaction, a baseline with climate damages needs to be compared to a hypothetical reference scenario in which climate change damages do not occur. These assumptions in the no-damage baseline are identical to those in the baseline projection, but exclude the economic consequences of climate change. Chateau et al. (2011) describe the baseline calibration procedure in more detail, although the numerical calibration of the model has since been updated to reflect more recent data. The no-damage baseline is a projection similar to the SSP2 standard scenario (Van Vuuren et al., 2014), but with revised socioeconomic drivers for population and economic growth. This "naïve" no-damage baseline projection, while purely hypothetical, provides the appropriate reference point for the analysis. It is differentiated from the core projection in which climate change impacts affect the economy, while all other assumptions remain unchanged.

2.1.1. *Macroeconomic activity and growth*

Demographic trends play a key role in determining long run economic growth. Projections of detailed movements in population by gender, age and education level determine future employment levels and human capital that drives labour productivity. While population and employment are correlated, the regional trends are differentiated by changes in participation rates for specific age groups (most prominently for people over the age of 65), changes in unemployment levels and changes in the age structure of the population (including aging).

Figure 2.1 presents the no-damage baseline projection of total regional population, based on the medium variant projection of the United Nations' World Population Prospects

Figure 2.1. **Trend in population by region, no-damage baseline projection**

Billion people

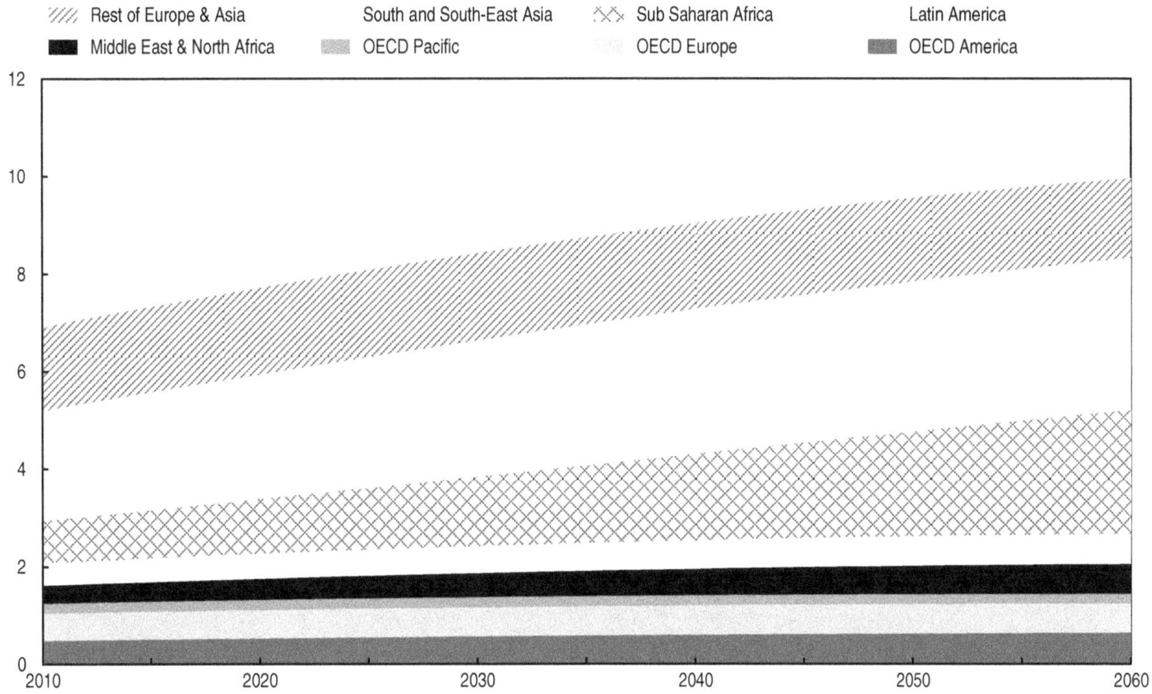

Source: UN (2013) as used in the ENV-Linkages model.

StatLink http://dx.doi.org/10.1787/888933275915

database (UN, 2013) and EUROSTAT (2013) for European countries.[2] At global level, population will increase from around 7 billion people in 2010 to almost 10 billion people in 2060. Despite the large increase, population growth by the middle of the century is projected to be substantially lower than it currently is. While this is true in most world regions, population keeps increasing at a steep rate in Sub-Saharan Africa.

The regional projections of GDP indicate that the slowdown in population growth does not imply a slowdown in economic activity. While long run economic growth rates are gradually declining, Figure 2.2 shows that GDP levels in the no-damage baseline are projected to increase more than linearly over time. The largest growth is observed outside the OECD, especially in Asia and Africa, where a huge economic growth potential exists. The share of the OECD in the world economy is projected to shrink from 64% in 2010 to 38% in 2060. These projections are fully aligned with the OECD Economic Outlook (OECD, 2014) and includes the main effects of the recent financial crisis as they emerged until 2013 and is consistent with the central scenario of the OECD@100 report on long-term scenarios (Braconier et al., 2014).

Besides labour supply, GDP growth is also influenced by changes in man-made capital and the use of land resources. In all cases, GDP growth is driven by a combination of increased supply of the production factors, changes in the allocation of resources across the economy, and improvements in the productivity of resource use (the efficiency of transforming production inputs into production outputs). Table 2.1 shows the average GDP growth rates for the current decade (2010-20), the medium term (2020-40) and the long term (2040-60). In most countries, short-term growth is primarily driven by a variety of sources, depending on the characteristics of the current economy. These short-term projections are based on the official forecasts made by OECD (2014) and IMF (2014). In the

Figure 2.2. **Trend in real GDP, no-damage baseline projection**

Billions of USD, 2005 PPP exchange rates

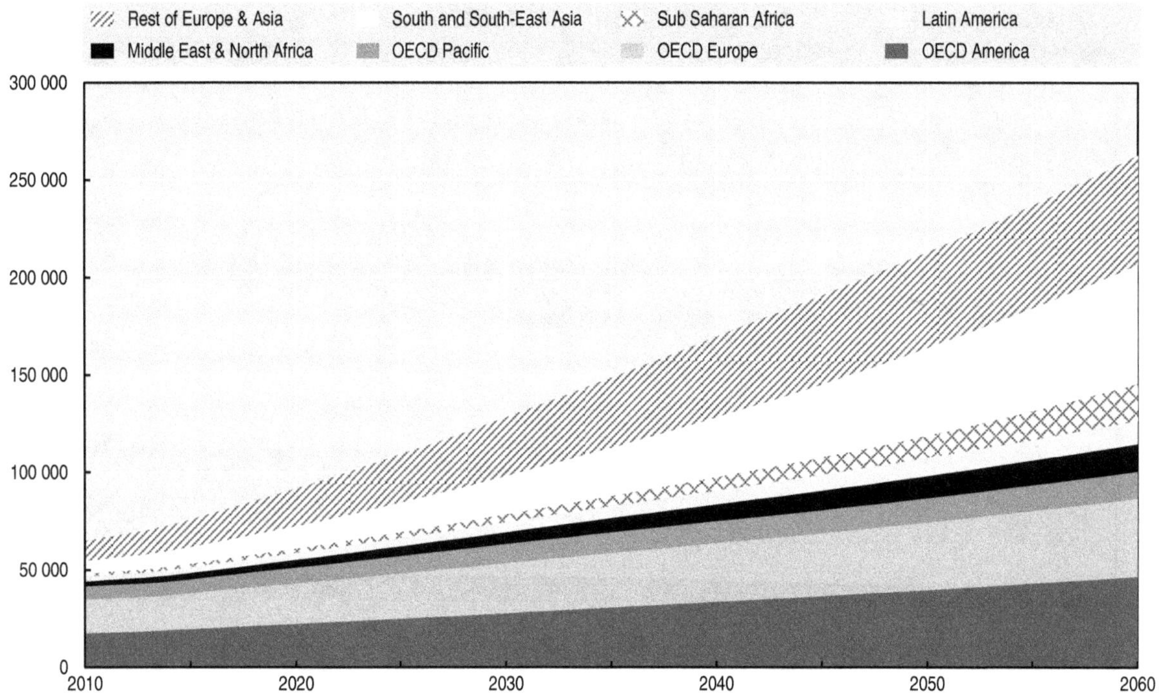

Source: OECD (2014) for OECD countries and ENV-Linkages model for non-OECD countries.

StatLink http://dx.doi.org/10.1787/888933275922

Table 2.1. **Economic growth over selected periods by region**

Average annual percentage GDP growth rates

	2010-20	2020-40	2040-60		2010-20	2020-40	2040-60
OECD America				Rest of Europe and Asia			
Canada	2.2	2.0	1.9	China	7.6	4.2	1.6
Chile	4.7	2.4	1.4	Non-OECD EU	2.2	2.5	1.7
Mexico	3.6	3.4	2.5	Russia	3.6	2.1	0.9
USA	2.4	1.9	1.5	Caspian region	6.3	4.8	2.6
OECD Europe				Other Europe	2.4	3.3	2.0
EU large 4	1.5	1.6	1.3	Latin America			
Other OECD EU	1.9	2.0	1.3	Brazil	3.3	3.0	1.8
Other OECD	3.6	2.6	1.7	Other Lat. Am.	3.6	3.7	3.1
OECD Pacific				Middle East and North Africa			
Aus. and New Z.	3.2	2.6	2.1	Middle East	3.4	3.7	2.3
Japan	0.9	1.0	1.1	North Africa	3.9	4.9	3.2
Korea	4.0	2.3	0.6	South and South-East Asia			
				ASEAN 9	4.8	4.2	3.1
				Indonesia	6.1	4.6	3.3
				India	6.6	5.8	3.6
				Other Asia	4.2	4.2	3.7
				Sub-Saharan Africa			
				South Africa	4.9	4.2	1.9
				Other Africa	5.9	6.5	6.0
OECD	2.2	1.9	1.5	World	3.5	3.1	2.2

Source: OECD (2014) for OECD countries and ENV-Linkages model for non-OECD countries.

StatLink http://dx.doi.org/10.1787/888933276218

longer run, a transition emerges towards a more balanced growth path in which labour productivity as a driver of economic growth is matched by increases in capital supply.

The table illustrates the main trends in economic development for the coming decades: continued slower growth in the OECD than in non-OECD countries (with a few exceptions), declining growth rates in emerging economies and relatively strong growth in Africa and most other developing countries.

2.1.2. *Structural change in the economy*

For an understanding of the future economy, it does not suffice to look at the macro economy only. To name just a few examples, projected productivity increases vary between different sectors, increasing incomes imply a change in demand for various goods, there will also be changes in the preferences of consumers, and international trade patterns may gradually adjust to stabilise trade balances.

Figure 2.3 shows how the sectoral structure in the OECD economies evolves, with the services sectors accounting for more than half of the GDP (i.e. value added) created in the future OECD economies. Generally, the shares of the various sectors in the economy tend to be relatively stable, although there are undoubtedly many fundamental changes at the sub-sectoral level that are not reflected here. The major oil exporters in the Middle East and Northern Africa are projected to gradually diversify their economies and rely less on energy resources. In developing countries the trend for a decline of the importance of agriculture is projected to continue strongly. Given the high growth rates in many of these economies, this does not mean an absolute decline of agricultural production, but rather an industrialisation process, and, in many cases, a strong increase in services.

Figure 2.3. **Sectoral composition of GDP by region, no-damage baseline projection**

Percentage of GDP

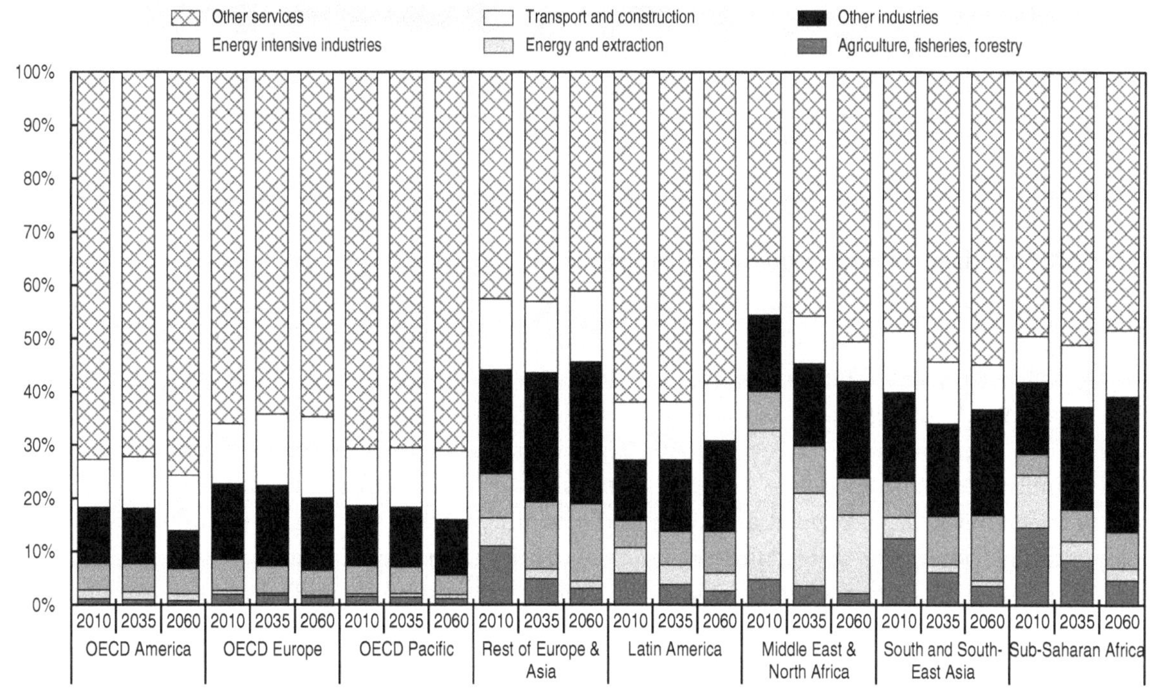

Source: ENV-Linkages model.

StatLink http://dx.doi.org/10.1787/888933275936

Energy projections until 2035 are calibrated to be in line with the Current Policies scenario of the International Energy Agency's World Energy Outlook (IEA, 2014), and extrapolated to fit the macroeconomic baseline thereafter. In fast-growing economies such as China, India and Indonesia, the need to support economic growth with cheap energy drives an increased use of coal, which is abundant and cheap in the absence of carbon pricing. In OECD regions, however, energy use is projected to switch towards more gas, not least in the United States. Furthermore, energy efficiency improvements dominate and imply a relative decoupling of energy use and economic growth. The resulting effects on energy production by fuel and region are given in Figure 2.4.

Figure 2.4. **Primary energy production, no-damage baseline projection**

Million tonnes of oil equivalent

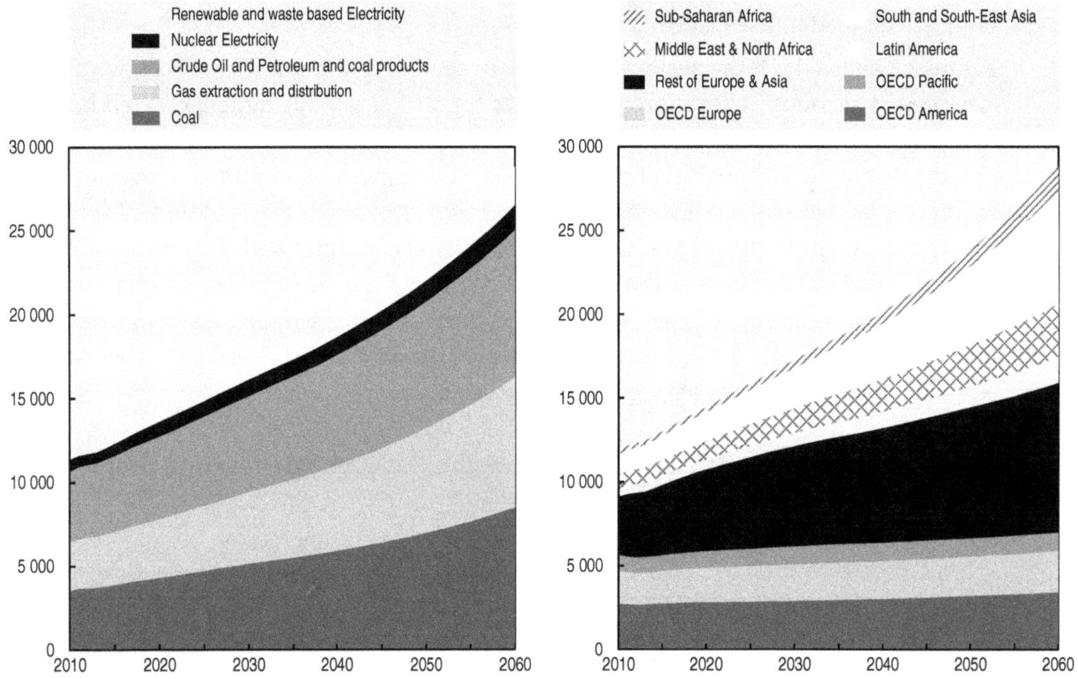

Source: ENV-Linkages model based on IEA (2014).

StatLink http://dx.doi.org/10.1787/888933275944

2.1.3. *Emissions and temperature increases*

Figure 2.5 illustrates how baseline economic activities lead to a steady increase in regional and global emissions. Global anthropogenic greenhouse gas (GHG) emissions (excl. emissions from land use, land-use change and forestry, which are treated exogenously) are projected to rise from around 45 Gigatonnes (Gt) of CO_2 equivalent (CO_2e) in 2010 to around 95 GtCO_2e in 2060. Carbon dioxide (CO_2) is projected to remain the dominant greenhouse gas. The rapid emission growth follows the key demographic projections of larger populations, increased economic activity and greater consumption of fossil fuel energy. Despite slowdowns in the growth rates of population and GDP, the shift in economic significance to emerging and developing economies, and – in the absence of new climate policies – unabated use of fossil fuels lead to a sharp increase in GHG emissions. In particular, the increased consumption of coal (as explained in the previous section) accelerates increases in emissions. Nonetheless, there is some relative decoupling: emissions grow less rapidly than production.

Figure 2.5. **Evolution of greenhouse gas emissions, no-damage baseline projection**

Gigatonnes of CO_2 equivalent

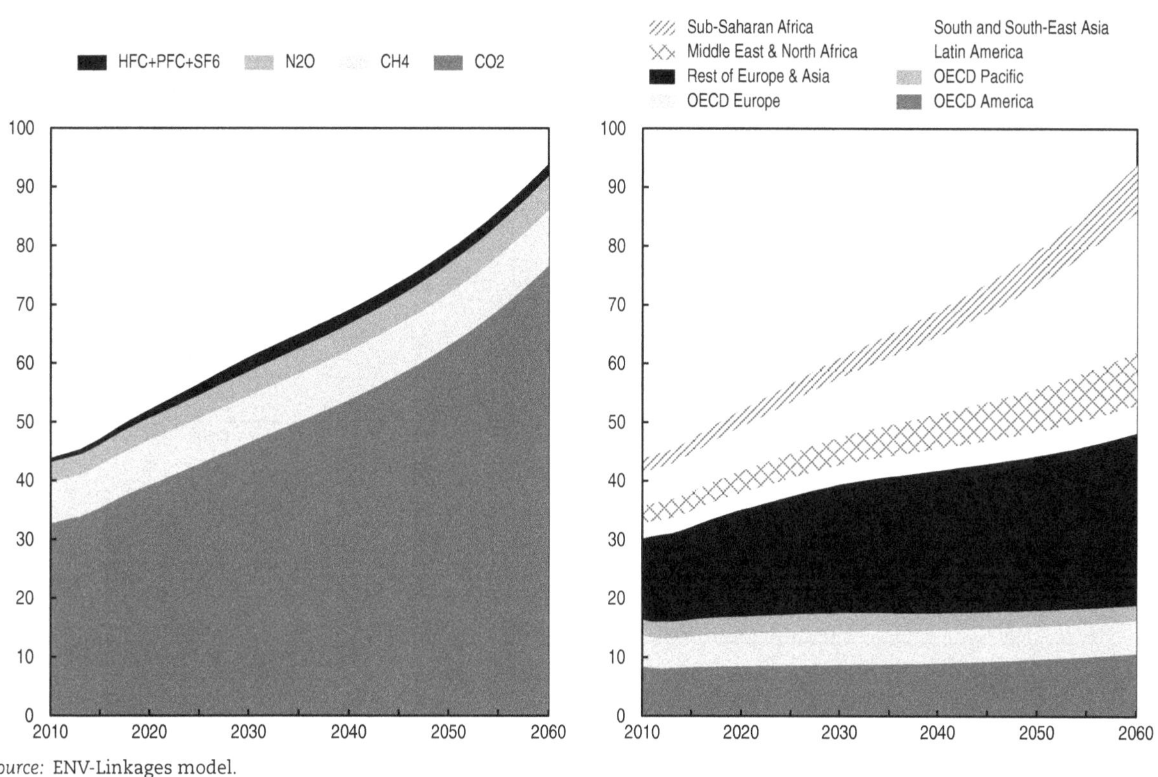

Source: ENV-Linkages model.

StatLink http://dx.doi.org/10.1787/888933275952

The rapid increase in GHG emissions accelerates climate change. Although the climate system is very complex and a whole range of biophysical processes are triggered by higher carbon concentrations (IPCC, 2014a), the focus of this report is on the economic consequences of climate change. Thus, only the main steps in the relation between economic activity and climate change are summarised: global concentrations from CO_2, and from the full basket of GHGs in CO_2 equivalents (Figure 2.6), radiative forcing (i.e. the change in the earth's radiation due to increased concentrations of GHGs) from anthropogenic sources (Figure 2.6) and global average temperature increases above pre-industrial levels (Figure 2.7). Concentrations of CO_2 in the atmosphere rise from 390 parts per million (ppm) to 590 ppm between 2010 and 2060. These concentration levels, plus forcing from other GHGs and aerosols lead to an increase in total radiative forcing from anthropogenic sources from just over 2 to almost 5 Watts per square meter (W/m^2).

There is substantial uncertainty on the temperature changes implied by these carbon concentrations and radiative forcing. The equilibrium climate sensitivity (ECS) reflects the equilibrium climate response, i.e. the long-run global average temperature increase, from a doubling in carbon concentrations, and is often used to represent the major uncertainties in the climate system in a stylised way. According to IPCC (2013), "ECS determines the eventual warming in response to stabilization of atmospheric composition on multi-century time scales". There are different ways to estimate ECS values, the most common being the use of instrumental climate system models or paleo-climatic observations. The central projection uses an ECS value of 3°C, even though the IPCC has not specified a median value. Where applicable, the ECS is varied between 1.5°C and 4.5°C in the likely

Figure 2.6. **Key climate indicators, no-damage baseline projection**

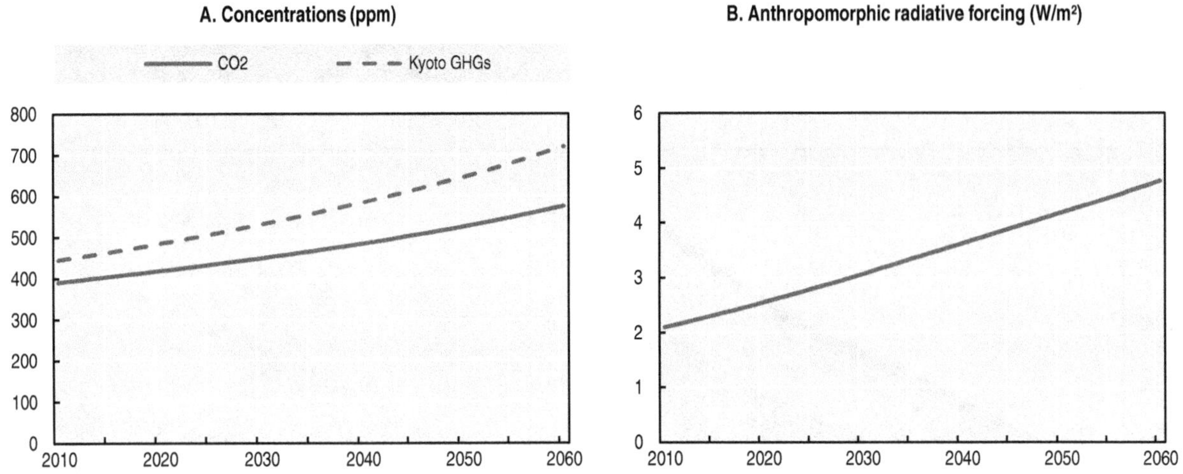

A. Concentrations (ppm)

CO2 — Kyoto GHGs

B. Anthropomorphic radiative forcing (W/m²)

Source: ENV-Linkages model and MAGICC6.4 (Meinshausen et al., 2011).

StatLink http://dx.doi.org/10.1787/888933275962

Figure 2.7. **Global average temperature increase, no-damage baseline projection**

Degrees Celsius above pre-industrial

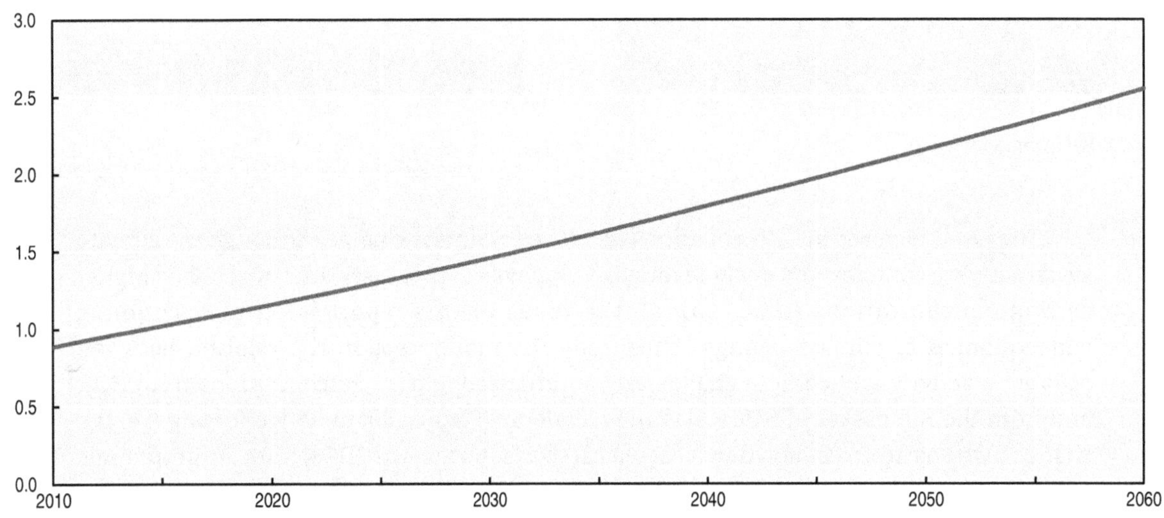

Source: ENV-Linkages model and MAGICC6.4 (Meinshausen et al., 2011).

StatLink http://dx.doi.org/10.1787/888933275975

uncertainty range, and between 1°C and 6°C in the wider uncertainty range, in line with the 5th Assessment Report of the Intergovernmental Panel on Climate Change (IPCC) (Rogelj et al., 2012; IPCC, 2013). The central projection delivers temperature increases of more than 2.5°C by 2060, as shown in Figure 2.7. This global temperature increase by 2060 is affected by the uncertainty on the ECS; the likely range equals 1.6 to 3.6°C, while the larger range is 1.1 to 4.3°C.

The regional impacts of climate change that are quantified in this study (cf. Chapter 1) are based on more detailed projections of regional changes in temperatures and precipitation patterns. The uncertainties on these regional patterns of climate change exist even for a given ECS, and are wider than the global average temperature change, but cannot be fully accounted for in the simulation of the economic damages. More elaborate robustness analysis, by varying the underlying climate model, and using results from a range of models for the climate system, crop yields and hydrology, is left for future research. The ISI-MIP project (Schellnhuber et al., 2014) provides some preliminary insights into the potential of using a multi-model comparison exercise to clarify various uncertainties.

2.2. Central macroeconomic and regional projections of damages from selected climate change impacts

2.2.1. Macroeconomic consequences

The selected regional impacts of climate change, as presented in Section 1.4, affect the economies of all countries in the world, but regional differences are substantial. While impacts become more severe over time, they are already measurable in the coming decades, indicating that the consequences of climate change are not just an issue for the distant future.[3] Perhaps more importantly, emissions between now and 2060 commit the world to a deteriorating risk profile: a high-emission infrastructure is locked in, damages continue for a century or more, and the risk of large-scale disruptions ("catastrophes") increases (see Chapter 1). Other impacts that are not included in the analysis but which potentially also have large economic consequences are discussed in Chapter 3. These constitute further sources of uncertainty, and while some sectors in some regions may be able to reap gains, on balance the costs of inaction presented in this chapter are likely to underestimate the full costs of climate change impacts (see Chapter 3 and IPCC, 2014a).

As Panel A of Figure 2.8 shows, the most economically vulnerable regions – for the selected impacts captured in the CGE model – are in Africa and Asia, with GDP losses in 2060 amounting to 3.3% for Middle-East and Northern Africa, 3.7% for South-and South-East Asia and 3.8% for Sub-Saharan Africa, respectively. These consist to a large extent of relatively poor, highly populated countries that do not have a high capacity to deal with significant negative impacts. Until around 2040, the impacts on these three regions are very similar, but in the last decade, they start diverging, with damages rapidly increasing especially in South and South-East Asia.

Impacts in Latin America (-1.5% by 2060) and Rest of Europe and Asia (which includes China and Russia; -2.1% GDP loss by 2060) are fairly close to the global average of 2.0% GDP loss. The economies of the OECD countries are, on balance, much less affected, with losses in 2060 amounting to -0.2%, -0.3% and -0.6% for OECD Europe, OECD Pacific and OECD America, respectively, not least because many of these countries lie in temperate climate zones.

The numbers presented above are of course central projections and Figure 2.8 – Panel A also shows the uncertainty bands associated with these regional projections due to the range of equilibrium climate sensitive. There are other sources of uncertainty as well, and in general regional projections tend to be much more uncertain.

At this level of aggregation, all regions are negatively affected by climate change. Damages are also projected to grow faster than global GDP. However, this should be seen against the perspective of GDP growth in the no-damage baseline projection: until 2060, the projections indicate that climate change will slow down economic growth somewhat (see also Section 2.2.3), but will not halt growth altogether. Furthermore, some of the largest impacts are in regions that have relatively high growth rates in the absence of climate change (compare Figure 2.8 with Table 2.1). Hence, a key consequence of climate change is that it slows the rate at which non-OECD economies catch up with income levels in OECD member countries, but – at least in these simulations – does not arrest it.

Figure 2.8. **Regional damages from selected climate change impacts, central projection**

Percentage change in regional GDP

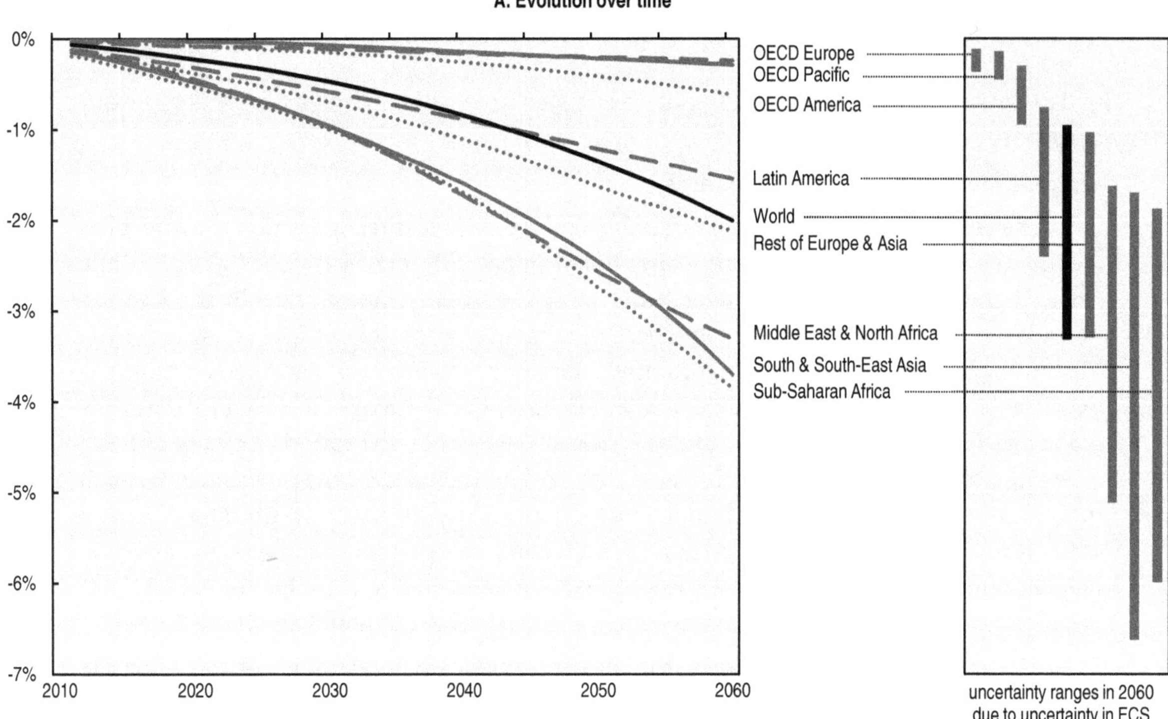

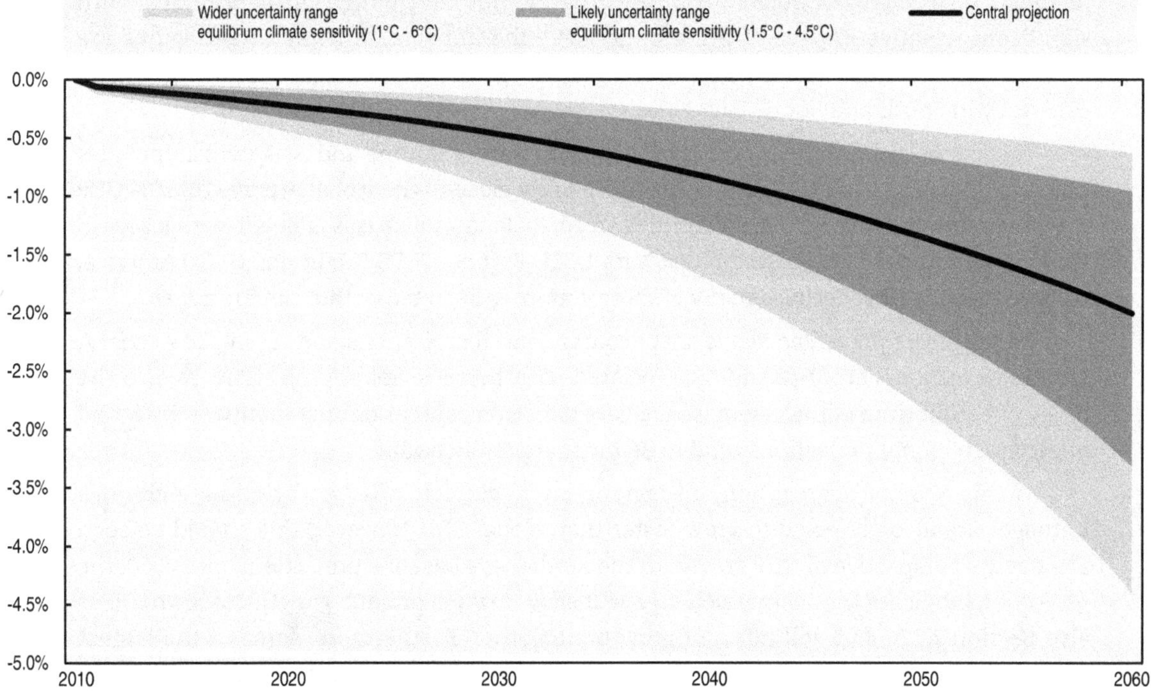

Source: ENV-Linkages model.

StatLink http://dx.doi.org/10.1787/888933275982

As the various climate impacts have different time profiles, the time profile of changes in GDP are also not identical across regions. The most vulnerable regions observe a nonlinear trend in damages as time (and climate change) progresses. This nonlinearity of regional and global damages over time reflects a combination of several key mechanisms. First, as regional temperatures increase, many of the negative impacts become more than proportionately stronger. Even though the global average temperature change projection is roughly linear (cf. Figure 2.7), biophysical impacts become more and more severe. This holds not least for yield losses in agriculture, where – especially in the temperate zones – small increases in temperatures do not affect (or even benefit) yields, while above a certain threshold yield losses rapidly increase. Second, nonlinearity arises from the limited possibilities in the economy to adjust to changes in factor supply and productivity. For small changes, reallocations within sectors and regions, and between regions, are relatively easy, but for larger shocks the reallocation costs become increasingly high. Third, losses are larger in regions that are projected to have high growth rates, and whose share in global GDP thus rises over time; this implies a shift in the weight of different regions in the global average and hence a non-linearity in global damages. Fourth, some of the impacts simulated here directly affect the growth rate of the economy, rather than merely the level of GDP. This holds especially for impacts that imply a destruction of the capital stock, such as coastal damages, which have permanent effects on the economy. While by 2060 all these individual mechanisms are rather small, they do add up to a clearly non-linear trend, and suggest that damages after 2060 may increase more and more rapidly.

Economic projections based on climate change impacts are, however, uncertain. The regional bands on the right-hand side of Panel A and Panel B of Figure 2.8 shows one element of the uncertainty (from equilibrium climate sensitivity) on these projections. The central projection, which leads to GDP losses that gradually increase to 2% in 2060, represents the central ("best guess") projection of damages as calculated with the available data for the subset of impacts studied. These projections are surrounded by a wide range of uncertainties in the economic and climate system, in the assessment of the climate impacts and in the way they feed back into the economy. As mentioned above, they also do not represent the full economic costs of climate change, as they do not include all market based impacts and exclude most elements of non-market impacts. Panel B illustrates a large uncertainty range corresponding to the equilibrium sensitivity of the climate system to accumulation of carbon (ECS). By 2060, annual GDP losses for the likely ECS range are 1.0% to 3.3%, but the possibility that global losses from the selected impacts covered in the model are as low as 0.6% or as high as 4.4% cannot be excluded. As this approximation of the impacts of uncertainty in the climate system is likely to be less robust at the sectoral level, ECS uncertainty ranges are only presented for the macroeconomic results. The regional uncertainties will be discussed in more detail below.

Due to lack of reliable data, a number of ad hoc assumptions underlie the calculations of the uncertainty range, not least the assumption that climate impacts scale proportionately with the value of the equilibrium climate sensitivity parameter. If damages increase more than proportionately in this parameter, the potential GDP losses will be larger than those shown in Figure 2.8. The uncertainty ranges given throughout this paper only reflect this particular – albeit important – uncertainty. Box 2.1 discusses how uncertainties are present at different stages; wider discussion of uncertainties is presented in Chapters 3 and 4.

The caveats on uncertainties in and incompleteness of these projections notwithstanding, the projections are well-aligned with the literature on quantified economic damages. The

Box 2.1. Main uncertainties in damage projections

Uncertainties can occur in every stage of the process to calculate damages from climate change (see Figure 1.1 for a simplified representation of these stages). A number of key uncertainties include:

1. Uncertainties in projecting the socioeconomic drivers of economic growth.
2. Uncertainties in projecting the mix of energy carriers used to produce energy.
3. Uncertainties in projecting the emission intensity of other emission sources (e.g. land use change emissions).
4. Uncertainties in projecting the climate system that links emissions to temperature change, including the ECS.
5. Uncertainties in projecting the regional patterns of climate change.
6. Uncertainties in projecting the impacts of climate change on specific impact categories.
7. Uncertainties in projecting the economic consequences of these impacts, including valuation of non-market impacts.

It is beyond the scope of this report to quantify each of these uncertainties, and they are not mutually independent. The IPCC reports discuss all of these uncertainties in detail, but do not provide a comprehensive overview of which type of uncertainty is largest. In its elaboration on the social cost of carbon (SCC), however, the IPCC highlights that "estimates can vary by at least two times depending on assumptions about future demographic conditions [...], at least approximately three times owing to the incorporation of uncertainty [...], and at least approximately four times owing to differences in discounting [...] or alternative damage functions [...]" (IPCC, 2014a). Despite the large uncertainties, the scientific evidence is overwhelming that the climate is changing as a result of GHG emissions. Thus, the uncertainties are on the size of the impacts and regional/local variations, not on the question whether climate change is real or not.

Wilby and Dessai (2010) offer an insightful conceptual review of uncertainty within the context of adaptation. A review of various sources of uncertainty in climate change economics is provided by Heal and Millner (2013), who stress that many uncertainties are irreducible and that uncertainties "explode" at finer spatial scales. Anthoff and Tol (2013) assess uncertainties by performing an extensive decomposition analysis using the FUND model. Watkiss and Hunt (2012) combine pairs of socio-economic scenarios, regional climate model outputs, temperature-related mortality functions, assumptions relating to acclimatisation, and valuation metrics with each other to project the economic impacts of climate change on human health. Large-scale multi-model intercomparison projects, such as ISI-MIP (Schellnhuber et al., 2014) shed further light on the size of various uncertainties. For example, Hinkel et al. (2014) apply the ISI-MIP philosophy of using multiple climate model runs and extensive sensitivity analysis to coastal flood damages and adaptation costs from sea level rise (but do not assess the macroeconomic consequences).

A separate strand of literature does not compare the uncertainties at different stages, but focuses on how uncertainties affect the Social Cost of Carbon (SCC). Otto et al. (2013), for example, discuss how specific climate system properties determine the SCC. Tol (2002; 2012) use econometric techniques to perform a quantitative meta-analysis of the literature to assess the central estimates and uncertainties on the SCC. EPRI (2014) dive into the details of the SCC values used for regulatory policies in the USA to assess whether uncertainties are properly captured in the existing estimates.

Box 2.1. **Main uncertainties in damage projections** *(cont.)*

In this report, the sensitivity analysis on the ECS tries to capture at least partially the uncertainties from stage 4. Some of the other uncertainties, not least in stages 5 through 7, are illustrated in an ad-hoc manner in Chapters 2 and 3 by presenting results for different climate models, and varying e.g. the crop model and the assumption on CO_2 fertilisation for agricultural impacts.

One aspect that is especially relevant for policy making is whether the economic uncertainties (stage 1) are larger than those of the climate system and impacts (stages 4 through 6). If the former is larger, it implies that those future projections where damages will be high are likely also those where incomes will be high. In this case, a risk-averse planner should not try to prevent such scenarios, but use a relatively high discount rate (Gollier, 2007, 2012; Nordhaus, 2013; Dietz et al., 2015); effectively, high economic uncertainty implies high potential benefits from mitigation action, to be discounted at high discount rates. If however the uncertainties in the later stages dominate, the appropriate discount rate will be lower and more ambitious immediate action is warranted to avoid potentially very bad (low-income, high-damage) outcomes. Naturally, the downside risks and uncertainties in the climate and impact assessments that are not related to the economic uncertainties still warrant a risk premium.

latest report of Working Group II of the Intergovernmental Panel on Climate Change (IPCC, 2014a) surveyed the existing literature and found "global aggregate economic losses between 0.2 and 2.0% of income ("medium evidence, medium agreement", Ch. 10) for a temperature increase of 2.5°C (this is not linked to a specific date). In the central projection of ENV-Linkages, this threshold is reached just before 2060. Given the wide range of impacts included in this analysis, it is not surprising that the GDP losses projected here are at the higher end of the range provided by the IPCC. Although there are significant differences between the modelling approach and calibration used here and earlier economic studies of climate damages (not least in the calibration of the impacts and the specification of the economic response through national and international substitution effects), similar patterns emerge in e.g. Nordhaus (2007; 2011), Eboli et al. (2010), Bosello et al. (2012), Roson and Van der Mensbrugghe (2012), Bosello and Parrado (2014) and Ciscar et al. (2014). In these studies, global impacts are increasing more than proportionately with temperature increases (and hence over time) and amount to reductions of several per cent of GDP by the end of the century. Highest impacts are foreseen in emerging and developing countries, especially in South and South-East Asia and Africa, whereas countries at a high latitude in the Northern hemisphere, especially Russia, may be able to reap some economic benefits from the climatic changes. Studies that focus on a specific region or impact, tend to show larger negative impacts on the local economy, but by nature ignore the endogenous adjustment processes that take place within economies, and changes in international trade patterns.

In comparison to the preliminary analysis, as documented in Dellink et al. (2014), the damages presented here are larger. Whereas the preliminary analysis had projected global losses of 0.7% to 2.5% of global GDP, this report finds losses in 2060 amounting to 1.0% to 3.3% GDP. There are two types of changes that affect this amended outcome. In the first place, a number of important climate impacts are added to the analysis, not least occupational heat stress and some aspects of extreme precipitation. All of these have negative consequences for the majority of countries. Secondly, the impacts on ecosystem services, which represented a willingness to pay valuation rather than a market impact, have been removed from the modelling exercise and included in Chapter 3 as a stand-alone assessment. This was done to more adequately reflect the nature of the projections on loss of biodiversity and ecosystems services, which are based on a willingness to pay valuation rather than an assessment of market damages. Moreover, the revision of the model to include 8 crop sectors rather than 1, and the change of the data source for the crop yield impacts, imply changes in the way economies respond to yield losses.

The economic consequences of climate change vary across the selected impacts that have been included in the modelling analysis, with the health impacts and agriculture (excluding a CO_2 fertilisation effect) contributing most to reduced economic growth. Figure 2.9 shows that of the impacts captured in the analysis, those related to health dominate in the current decade. This is dominated by labour productivity losses from occupational heat stress, even though this excludes the welfare losses from premature deaths (see Chapter 3). By 2060, the health impacts contribute to 0.9% GDP loss in the central projection. This dominance of labour productivity losses from occupational heat stress is also found in e.g. Roson and Van der Mensbrugghe (2012) and Ciscar et al. (2014), while studies that show a dominance of agricultural impacts tend to exclude heat stress effects (e.g. Bosello et al., 2012; Bosello and Parrado, 2014).

Figure 2.9. **Attribution of damages to selected climate change impacts, central projection**

Percentage change in GDP w.r.t. no-damage baseline

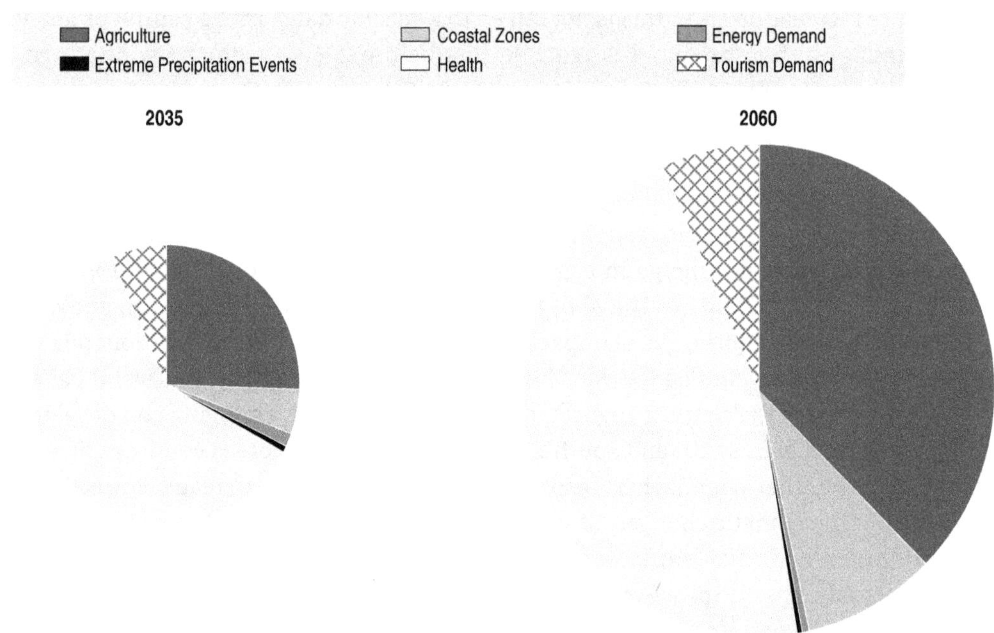

Source: ENV-Linkages model; see Table 1.1 in Chapter 1 for which impacts are modelled in each category.

StatLink http://dx.doi.org/10.1787/888933275998

Over time, agriculture emerges as a second major impact category (at least when excluding the effects of CO_2 fertilisation, as is the case in the central projection): between 2040 and 2060, the health damages less than double, while the agricultural damages increase threefold. By 2060, the agricultural damages (including those to fisheries) contribute 0.8%

GDP loss. Although in most regions the share of agriculture in GDP is modest, the yield impacts are projected to be quite significant (cf. Rosenzweig et al., 2013). This projection on agricultural damages is subjected to a sensitivity analysis in Section 2.2.5, by varying the assumptions underlying the impacts in terms of the crop model that is used to project yield shocks, the climate model used for regional patterns of climate change, and – not least – the effect of CO_2 fertilisation. As global food security increases significantly in the no-damage baseline, overall food security may not be threatened too much if trade patterns are flexible enough to accommodate changes in regional productivities, although a detailed assessment of food security is beyond the scope of this report.

Economic damages to coastal zones and tourism demand – for the selected impacts considered in the CGE model – are smaller, contributing to 0.2% and 0.1% GDP loss, respectively. It is stressed that these are not the full economic costs of these sectors – see discussion in the next chapter. Especially for coastal zones, the impacts are likely to be felt more strongly in later decades; between 2040 and 2060, the increase in coastal zone damages even outstrips those of agriculture. The global damages from energy are very small, as there are offsetting effects from reduced heating and increased cooling. Finally, the projected economic consequences of extreme events (from hurricanes) are negligible at the global level, but this is at least partially caused by the absence of flood damages in the quantitative assessment; furthermore, damages from heat stress on humans, which are significant, are accounted for in the health impact category, while heat stress for crops is included in the impacts on agriculture.

2.2.2. Regional consequences

Given the large regional differences in climate change impacts, and the differences in economic structures of economies, a more detailed regional analysis can shed more light on where the economic consequences of the selected impacts climate change are the largest. Although uncertainties on regional changes in the climate system are larger than at the global level, the analysis can be extended to the level of the 25 regions of the ENV-Linkages model. Figure 2.10, Panel A illustrates that within the large macro-regions, there are substantial differences in climate impacts. Of the 25 regions presented in Panel A, 22 are negatively affected by climate change, representing 94% of the current world economy and 97% of current global population,[4] and all regions are negatively affected by at least some climate impacts.

For most countries and regions, the economic consequences of climate change will be clearly negative. In line with the global results, impacts on health and agriculture tend to dominate in the modelling analysis, although by 2060 many vulnerable countries also suffer from the impact on coastal zones of sea-level rise, and it is likely that the effect of this particular impact category becomes even more pronounced after 2060. The regional impacts on tourism demand fluctuate more across regions than the other impacts, with benefits in some, and large losses in other countries. This is partly due to the nature of the biophysical impacts, especially their dependence on regional climatic changes, but is also driven by the international trade links between countries. For instance, some of the main agricultural crops are heavily traded and small changes in productivity can imply relatively large shifts in trade – and hence production – patterns. In most regions a large share of damages comes from agriculture. This is however heavily dependent on the initial assumptions made about agricultural impacts, which are surrounded by large uncertainties. Section 2.2.5 explores this with a sensitivity analysis on the agricultural impacts.

Figure 2.10. **Damages from selected climate change impacts, central projection**

Percentage change in GDP in 2060 w.r.t. no-damage baseline

A. Damages by region and impact category in 2060

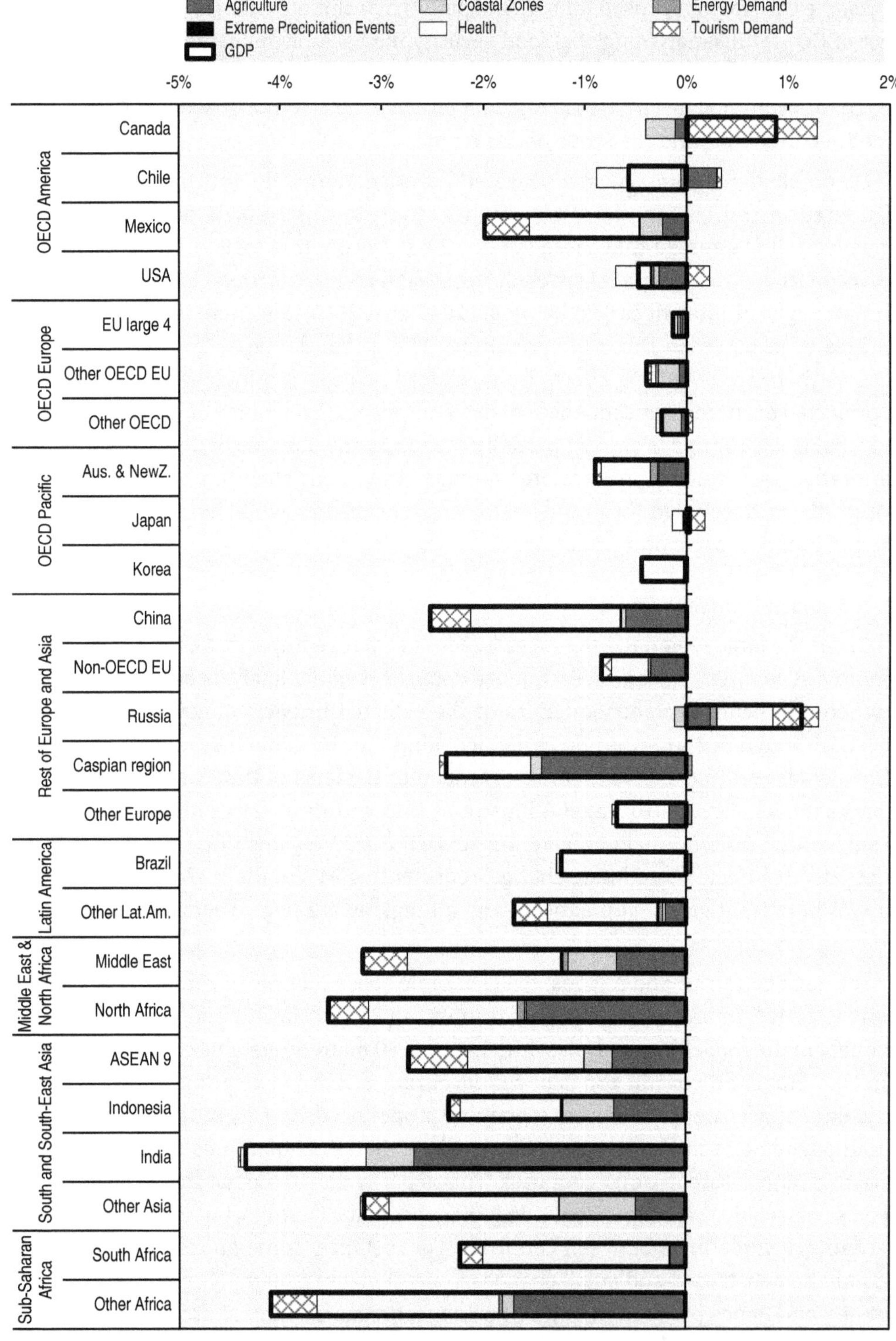

Figure 2.10. **Damages from selected climate change impacts, central projection** *(cont.)*

B. Uncertainty in regional damages in 2060 due to uncertainty in equilibrium climate sensitivity

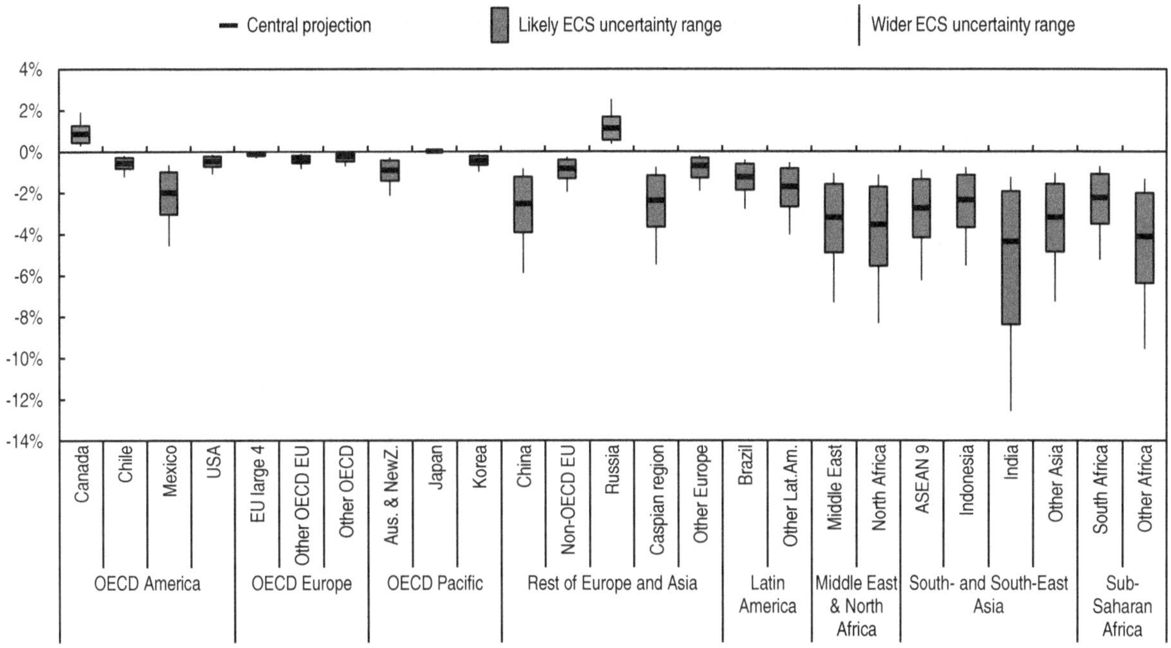

Central projection | Likely ECS uncertainty range | Wider ECS uncertainty range

4% 2% 0% -2% -4% -6% -8% -10% -12% -14%

Canada, Chile, Mexico, USA — OECD America; EU large 4, Other OECD EU, Other OECD — OECD Europe; Aus. & NewZ., Japan, Korea — OECD Pacific; China, Non-OECD EU, Russia, Caspian region, Other Europe — Rest of Europe and Asia; Brazil, Other Lat.Am. — Latin America; Middle East, North Africa — Middle East & North Africa; ASEAN 9, Indonesia, India, Other Asia — South- and South-East Asia; South Africa, Other Africa — Sub-Saharan Africa

Note on Panel A: The black rectangles represent total GDP impact (central projection).
Source: ENV-Linkages model.

StatLink http://dx.doi.org/10.1787/888933276005

The most severely affected regions are India, Sub-Saharan Africa (excluding South Africa), the countries in the Middle East and Northern Africa and other developing Asian economies. In India, agricultural losses are especially large (with large reductions in value added created in all crop sectors), but impacts on energy and tourism demand hardly play a role. Although the literature stresses especially the negative consequences of increased demand for cooling (Akpinar-Ferrand and Singh, 2010; Mima et al., 2011), the projections show that reduced heating expenditures largely compensate for this, especially in emerging economies where the baseline (following the IEA World Energy Outlook) projects a significant increase in energy demand for heating in combination with a trend towards further electrification of the heating systems. Box 2.2 shows that qualitatively, the very small energy damages are rather robust, by comparing with projections using an alternative energy model to specify changes in demand by energy carrier (but without changing the baseline assumptions on the evolution of energy demand in absence of climate change). In Sub-Saharan Africa, strong agricultural impacts are coupled with productivity losses from heat stress, increased energy use for cooling and a smaller increase in tourism revenues than would be realised without climate change to explain the significant GDP losses. In the Middle East and North Africa, these categories are also important, but complemented by substantial coastal zone impacts from sea level rise, which generally play a much larger role in Asia than for economies on other continents as people and capital in Asia tend to be more concentrated in coastal areas.

In the 25 region aggregation, Canada and Russia are the only two economies that are projected to experience positive net impacts from climate change by 2060, at least for the impacts included in the modelling analysis. Logically, net benefits may also be expected for other countries with similar climate conditions, but these are hidden in larger regional groups; for instance, the Other OECD EU region covers both Scandinavian and Mediterranean countries, and impacts at the national level may vary widely from the regional group's average. That of course does not mean there are no climate impacts in these countries; rather, the positive monetary consequences of the selected impacts covered in the model outweigh the negative ones and the main net impacts in this region are non-monetary, as discussed in Chapter 3. For Canada, the main benefit are projected to come from beneficial impacts on tourism. For Russia, the benefits are more diverse, and also include agriculture, energy and health. Minor benefits from climate change may also arise in other regions, e.g. agriculture in Chile (where positive impacts on fruit and vegetables outweigh negative impacts on wheat).

The modelling also suggests minor net economic benefits from health impacts in some European Union countries (where relatively small losses in labour productivity are combined with slightly beneficial impacts on health expenditures and disease incidence), so that on balance the GDP of the large 4 EU countries (as a group, rather than individually) is not affected at all by the selected climate change impacts assessed. The more detailed Peseta II study on impacts in Europe carried out by the EU's Joint Research Centre (Ciscar et al., 2014) reveals that there are sizable differences within Europe, with negative consequences of occupational heat stress concentrated in the southern European countries. In most cases, the gains do not reflect more beneficial climate circumstances, but rather an improvement in the relative competitive position of certain sectors in these economies vis-a-vis their main competitors. Such trade implications of climate impacts will be investigated in more detail in Section 2.2.4.

In most regions, the macroeconomic consequences of the interactions between the various impacts are very small. In these cases, the total GDP loss is very close to the sum of the losses from the individual impacts. This is logical, as the various impacts affect different parts of the economy, and the overall damages are relatively small, so that interaction effects are minor. For more severely affected countries, especially India, these interaction effects are non-negligible, and the total GDP loss is smaller than the sum of the individual losses, indicating that countries can respond to the variety of different impacts in a more sophisticated way than simply responding to individual impacts.

Panel B of Figure 2.10 presents the associated uncertainty levels from varying the ECS. The blue bars indicate how much regional damages may fluctuate in the likely range (1.5°C-4.5°C), while the thin black lines highlight that the impacts may be considerably larger (or smaller) when an even wider uncertainty range (1°C-6°C) is considered. The panel clearly shows that the regional differences in GDP losses in the central projection are relatively small when compared to the uncertainties within a region related to different climate sensitivities. Perhaps equally importantly, the model analysis shows that even at low levels of climate sensitivity there will be significant (albeit smaller) GDP losses in many countries. Within the OECD America group, the largest uncertainties related to the equilibrium climate sensitivity are in Mexico, with losses potentially exceeding 4% by 2060; in contrast, the ranges for e.g. the United States are much smaller. For the OECD America group as a whole (not presented in the figure), the likely range of damages is 0.3% to 0.9% of GDP. For OECD Europe, where projected impacts are smaller, the likely range is also smaller: 0.1% to 0.4% of GDP by 2060 for the group. For the Asian and Pacific OECD countries, the likely range is larger in the pacific member countries than for the Asian members, which leads to a likely range of 0.1% to 0.5% of GDP by 2060. Of course, this excludes other uncertainties, which may have significantly different regional patterns (cf. Box 2.1).

For the non-OECD regions, the variations are often substantially larger. The potential GDP losses in India and Sub-Saharan Africa when climate sensitivity is higher than the central projection are large, and could run into the double digits for (very) high values of the climate sensitivity. The likely range for India is 1.9% to 8.4% of GDP by 2060, and for Sub-Saharan Africa 2.0% to 6.4%. Thus, while the central projection for both is rather close, the possible negative consequences for India are projected to be substantially larger.

Box 2.2. **Sensitivity to the energy impacts**

Several earlier studies have used the POLES model to project the impacts of climate change on energy demand (Criqui, 2001; Criqui et al., 2009), which was also used in the EU ClimateCost project (Mima et al., 2011). POLES is a bottom-up partial-equilibrium model of the world energy system. Similar to the IEA's WEM model used in the central projection, it determines future energy demand and supply for coal, oil, natural gas, electricity according to exogenous trends and climate impacts through their effects on the number of days that require heating or cooling (heating degree and cooling degree days). The projected impacts of climate change, using the SRES A1B scenario (which is similar to the CIRCLE baseline), differ substantially between both models, not least due to different assumptions on the trends in the no-climate change baseline. Particularly, Poles projects much smaller impacts of changes in heating on electricity demand, and hence projects an increase in electricity demand for most regions. In line with the WEM projections, changes in demand for fossil fuels are relatively small. Table 2.2 shows how according to the ENV-Linkages simulations these alternative assumptions affect the consequences for GDP in the various regions, excluding the other categories of climate impacts, i.e. focusing purely on the impacts of changes in energy demand. What is clear from the table is that despite substantial differences in underlying impacts, the macroeconomic consequences are projected to be very small (a conclusion that is shared with Mima et al., 2011), with the largest effects in the Middle East. The two models disagree on the impacts in South and South-East Asia: while in the IEA projections the reduced demand for heating (which increases rapidly in the baseline) dominate, the Poles projections are dominated by increased demand for cooling. IEA (2013) stress that impacts on energy supply may lead to higher macroeconomic damages.

Table 2.2. **Influence of alternative assumptions on energy impacts**

Percentage change in GDP in 2060 w.r.t. no-damage baseline

	2035		2060	
	Central projection %	Projection using Poles %	Central projection %	Projection using Poles %
OECD America	-0.01	0.00	-0.02	0.01
OECD Europe	-0.02	0.00	-0.02	-0.01
OECD Pacific	-0.01	0.00	-0.03	0.00
Rest of Europe and Asia	-0.05	-0.02	-0.06	-0.09
Latin America	-0.01	0.00	-0.02	0.00
Middle East and North Africa	-0.01	0.01	-0.03	0.01
South and South-East Asia	0.00	-0.01	0.01	-0.03
Sub-Saharan Africa	0.00	0.01	0.01	0.00
World	-0.01	0.00	-0.01	-0.01

Source: ENV-Linkages model based on IEA and Poles projections.

StatLink http://dx.doi.org/10.1787/888933276229

Although many of the most severely hit countries currently have a below average per capita income, the no-damage scenario also projects relatively high income growth rates in these countries (cf. Section 2.1). Hence, even when the impacts of climate change are factored in, the projections still show a substantial convergence of incomes across the globe, at least in percentage terms. Table 2.3 shows that absolute per capita consumption levels remain widely diverging, with or without climate damages. In most cases, the climate damages do not affect the projected ranking of countries by consumption level in 2060. Such average per capita consumption levels of course hide the substantial differences that impacts may have on the various household groups. In particular, it is warranted to further investigate to what extent the economic consequences of climate change are disproportionately felt by specific household groups. For instance, as food prices are projected to be more affected than most other commodity prices, agricultural impacts may pose more significant problems for poor households that are net food consumers in certain regions, while poverty rates may actually drop in other regions where many poor households tend to be net food producers, at least for mild levels of climate change (Hertel et al., 2010).

Table 2.3. **Per capita consumption levels over selected periods by region**

Thousands of USD, 2005 PPP exchange rates
Percentage change between both projections in parentheses

	2010	2035		2060	
		No-damage projection	Central projection	No-damage projection	Central projection
OECD America	24.9	34.6	34.5 (-0.1%)	47.2	47.0 (-0.5%)
OECD Europe	18.1	27.4	27.4 (-0.1%)	39.6	39.5 (-0.3%)
OECD Pacific	21.4	32.5	32.4 (-0.2%)	47.6	47.3 (-0.7%)
Rest of Europe and Asia	1.5	6.3	6.3 (-1.0%)	10.5	10.2 (-2.8%)
Latin America	3.3	6.9	6.8 (-0.8%)	11.6	11.3 (-2.1%)
Middle East and North Africa	2.0	4.6	4.5 (-1.5%)	9.4	9.0 (-4.0%)
South and South-East Asia	0.8	2.6	2.5 (-1.4%)	5.1	4.9 (-4.1%)
Sub-Saharan Africa	0.6	1.3	1.3 (-1.4%)	2.4	2.3 (-4.6%)
World	**4.9**	**8.0**	**8.0 (-0.5%)**	**11.6**	**11.4 (-1.8%)**

Source: ENV-Linkages model.

StatLink http://dx.doi.org/10.1787/888933276230

2.2.3. Consequences for the structure of the economy

Climate damages affect the economic structure of the various regions, as the various impacts are linked to inputs in different sectors. This is reflected by changes in the sectoral contributions to GDP. Figure 2.11 illustrates changes in regional value added in 2060 for the main aggregate sectors of the economy. Some sectors are directly impacted by specific climate impacts (e.g. services sectors are affected by health impacts, energy sectors by energy demand impacts, etc.). However, there are also substantial indirect impacts, such as changes in production in (energy-intensive) industrial sectors due to the full range of price changes that follow climate impacts or capital destruction from sea-level rise which affects all sectors through changes in the marginal productivity of capital.

Given the sheer size of the services sector in most economies, accounting for more than 50% of total GDP globally (both currently and in the coming decades), and more than that in the OECD countries, it is no surprise that this sector also takes on a large share in the changes in GDP levels from the no-damage baseline projection. This does not imply that the services sector is most severely hit by climate change – although some impacts do

Figure 2.11. **Sectoral composition of damages from selected climate change impacts, central projection**

Percentage change in GDP in 2060 w.r.t. no-damage baseline

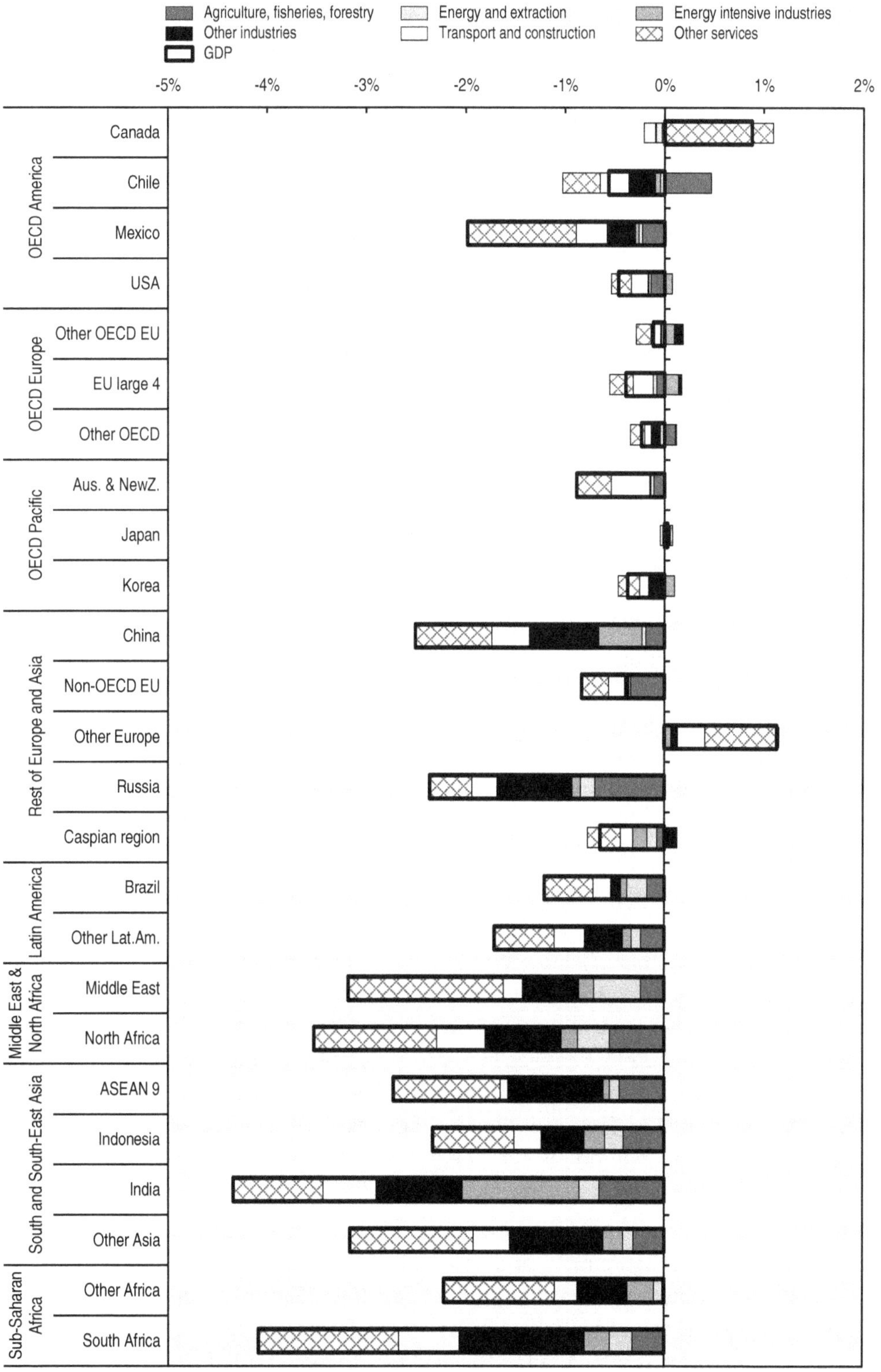

Agriculture, fisheries, forestry; Energy and extraction; Energy intensive industries; Other industries; Transport and construction; Other services; GDP

-5% -4% -3% -2% -1% 0% 1% 2%

OECD America: Canada, Chile, Mexico, USA
OECD Europe: Other OECD EU, EU large 4, Other OECD
OECD Pacific: Aus. & NewZ., Japan, Korea
Rest of Europe and Asia: China, Non-OECD EU, Other Europe, Russia, Caspian region
Latin America: Brazil, Other Lat.Am.
Middle East & North Africa: Middle East, North Africa
South and South-East Asia: ASEAN 9, Indonesia, India, Other Asia
Sub-Saharan Africa: Other Africa, South Africa

Note: The black rectangles represent total GDP impact (central projection).
Source: ENV-Linkages model.

StatLink http://dx.doi.org/10.1787/888933276015

affect this sector directly – but rather that the economic consequences of climate change spread throughout the economy. Thus, sector-specific studies, which focus on the consequences of climate change on a specific sector or group of sectors, can shed light on the direct consequences of climate change, but fail to account for the strong indirect effects occurring as impacts propagate through markets to other sectors.

In the countries that are most severely impacted by climate change, the production and value-added of all sectors is diminished. In these cases, all major sectors of the economy are directly affected, and there is less room for adjusting economic structures to accommodate production losses in specific sectors. Domestic cost increases from climate change damage imply an increase in the terms of trade of these countries, which is the defined as the relative price of exports in terms of imports. Yet, as export volumes decrease in these countries due to decreased global demand, the most vulnerable countries are not projected to increase revenue from exports.

The countries in OECD Europe are typical cases where very small macroeconomic impacts hide a more pronounced effect in specific sectors: trade-exposed industries can benefit from improved international trade, whereas the more sheltered services sectors are hurt by domestic tourism and health impacts, but also by reduced availability of capital from coastal damages.

Despite the relatively small contribution of the agricultural sectors to GDP in most countries, the contribution of agriculture to overall GDP loss is substantial, as this sector is projected to be substantially affected by climate change (the drop in global agricultural value added amounts to 9% by 2060). The endogenous economic response of the model ensures that for instance the reduction in food consumption is limited by intensifying agricultural production and allocating more land to crop production. Furthermore, these sector-specific impacts cause a re-allocation of production factors (and especially capital) across all sectors. The indirect impacts on other sectors show that adaptation through trade and production adjustments are a powerful instrument to reduce the economy-wide costs of climate change. The significance of trade as an adjustment channel is discussed further below in Section 2.2.4.

The GDP impacts can also be attributed to specific production factors. Climate impacts directly affect labour, capital, land and natural resources, as explained in Chapter 1. Furthermore, the economic adjustment processes also result in changes in the contribution of tax revenues to GDP, even though there are no direct impacts of climate change on taxes. Figure 2.12 shows the decomposition of GDP losses according to production factor, with shading indicating the direct changes in value added of a production factor. These direct effects have been calculated by multiplying the percentage change in productivity and supply of these production factors at their no-damage baseline levels of use, i.e. before any endogenous market adaptation effects. The indirect effects (not shaded in Figure 2.12) are then calculated as the difference between the total effect and the direct effect.

In the model, labour supply is assumed to be fixed, and land and natural resources are relatively inflexible in their supply, and hence direct effects more or less directly translate into GDP loses, although sectoral reallocation can still affect their overall contribution to GDP. An exception is the reduction of value added from natural resources in South and South-East Asia, which is attributed to the decline in production of a number of resource-dependant sectors, which is induced by the changes elsewhere in the economy.

Figure 2.12. **Sources of damages from selected climate change impacts by production factor, central projection**

Percentage change in GDP in 2060 w.r.t. no-damage baseline

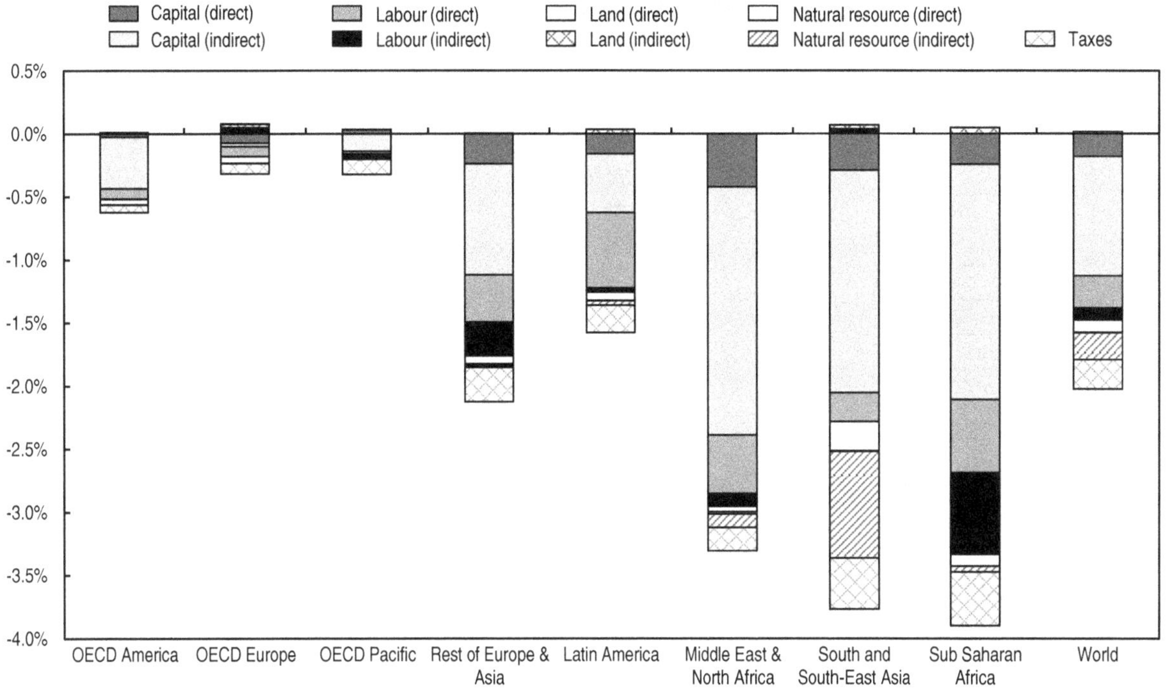

Capital (direct) Capital (indirect) Labour (direct) Labour (indirect) Land (direct) Land (indirect) Natural resource (direct) Natural resource (indirect) Taxes

0.5% 0.0% -0.5% -1.0% -1.5% -2.0% -2.5% -3.0% -3.5% -4.0%

OECD America OECD Europe OECD Pacific Rest of Europe & Asia Latin America Middle East & North Africa South and South-East Asia Sub Saharan Africa World

Source: ENV-Linkages model.

StatLink http://dx.doi.org/10.1787/888933276027

For capital, the situation is different, as supply is flexible in the long run, as consumers can adjust their savings patterns. Thus, there is an additional effect, as changes in income levels affect savings and hence future capital accumulation. Thus, the climate impacts not only affect the level of GDP, but also the growth rate, through reduced capital accumulation. As can be inferred from Figure 2.12, capital losses are substantially larger than the other factor losses, and this can be attributed to these indirect economic effects. At the global level, almost half of the projected GDP loss of 2% can be attributed to the indirect effects on capital, which may be interpreted as growth effects. In other words, by 2060 the projected economic consequences on GDP levels and on GDP growth are of similar size.[5]

2.2.4. Consequences for international trade

Economies do not operate in isolation, and the climate impacts in one region affect the economy in other regions as well. Figure 2.13 shows how different regions are affected differently by the adjustments in international links. The role of international links between regions can be analysed by comparing global climate impacts to the hypothetical case in which impacts occur only in each single region ("domestic impacts" in Figure 2.13). In the "unilateral" case, world market prices are not or hardly affected and thus domestic impacts dominate indirect effects through adjusting international trade patterns. In contrast, in the multilateral case the climate damages in all regions affect international trade patterns. If impacts were identical across countries, all regions would benefit from maintaining their international competitive position when the damages are global. Simultaneously, they would

Figure 2.13. **Domestic and global climate change damages from selected climate change impacts, central projection**

Percentage change in GDP in 2060 w.r.t. no-damage baseline

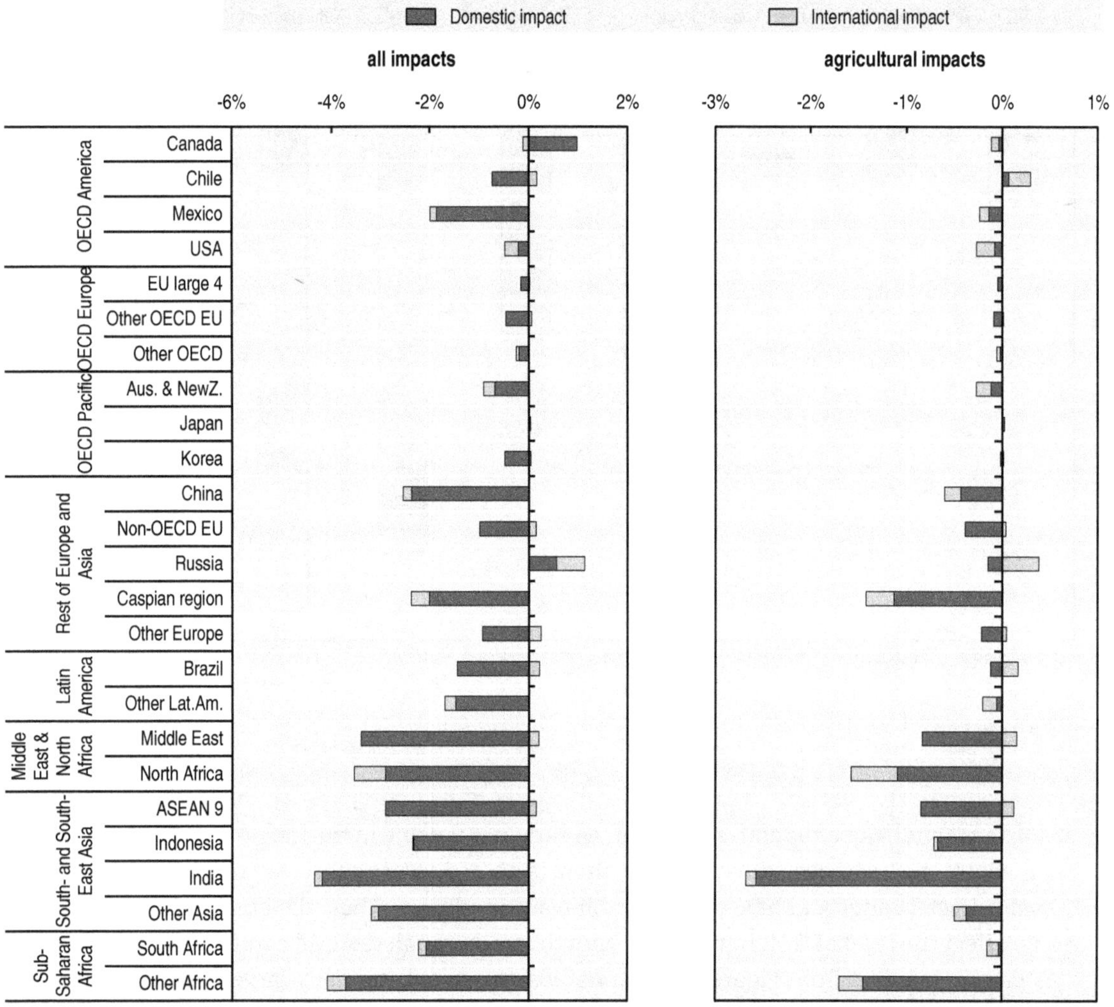

Domestic impact
International impact

all impacts
agricultural impacts

-6% -4% -2% 0% 2%
-3% -2% -1% 0% 1%

OECD America: Canada, Chile, Mexico, USA
OECD Europe: EU large 4, Other OECD EU, Other OECD
OECD Pacific: Aus. & NewZ., Japan, Korea
Rest of Europe and Asia: China, Non-OECD EU, Russia, Caspian region, Other Europe
Latin America: Brazil, Other Lat.Am.
Middle East & North Africa: Middle East, North Africa
South- and South-East Asia: ASEAN 9, Indonesia, India, Other Asia
Sub-Saharan Africa: South Africa, Other Africa

Note: The results in the right panel show the effect on total GDP, not just the value added generated in the agricultural sector.
Source: ENV-Linkages model.

StatLink http://dx.doi.org/10.1787/888933276032

be negatively affected by the reduced demand for exports following the economic slowdown in the trading partners who are affected by climate change. However, heterogeneity in impacts means that relative competitive positions start to shift, and if climate change is beneficial (or less negative) for the main trading partners, whereas the domestic damages are (more) negative, the gap in competitive position may further widen due to "international" damages. Together, these "domestic" and "international" damages determine the total damages, which correspond to the damages presented above in Figure 2.10.

In the left panel of Figure 2.13, the breakdown of domestic and international damages is shown for all impacts together (i.e. the central projection discussed above), while the right panel shows the results when only agricultural damages are considered. When considering all impacts, the international trade effects are positive for some and negative for other regions. At the global level, the international impacts are relatively small but negative. Damages in other regions will, on the one hand, negatively affect the domestic economy, especially because countries cannot protect their consumption levels by importing more from unaffected regions. On the other hand, countries are better able to maintain their regional competitive position when other regions are also affected, at least when the major trading partners are affected in a similar way. In the model simulations, the negative first effect dominates.

There are a few countries where large domestic impacts can be partially compensated by improved trade relations, including Chile, Brazil, the Middle East and the ASEAN9 economies. These tend to be relatively open economies. In contrast, in other regions, especially India and the African countries, but also e.g. the United States, the domestic effects are further aggravated by the loss in competitive position compared to their main trading partners. In turn, this leads to a further slow-down of these economies, lowering investment and consequently GDP levels. Effectively, more closed economies have to absorb a larger part of the shock domestically.

The gain in competitive position for some of the less affected countries occurs especially when the main trading partners, not least the major emerging economies in Africa and Asia, observe substantial losses in agricultural productivity. As the agricultural sector is especially trade-exposed, it is not surprising to see in the right panel of Figure 2.13 that the international impacts are relatively larger when only agricultural impacts are considered. The panel makes clear that most of the trade gains projected for the central projection with all impacts come from gains linked to agricultural damages in other regions. Domestic impacts do not need to be strictly beneficial to gain from international trade, as e.g. the projections for Russia, Brazil and the ASEAN9 region indicate. Rather, international trade gains are about *relative* competitive positions, not absolute. For the most severely affected regions the effect is opposite: their domestic damages are worsened by a loss of income from exports.

2.2.5. A sensitivity analysis on agricultural damages

The economic consequences of the agricultural impacts described above are based on projections made within the AgMIP project and exclude the effects of CO_2 fertilisation (Nelson et al., 2014; Von Lampe et al., 2014; see also Chapter 1). One specific scenario was chosen for the central projection, using projections of regional climate changes from the HadGEM model and using the DSSAT crop model to identify the agronomic impacts. In the AgMIP project, alternative choices are available for both the climate system model (using IPSL instead of HadGEM) and the crop model (using LPJmL instead of DSSAT), which are used here to illustrate the uncertainties.[6] As a sensitivity analysis, alternative scenarios are analysed in this section, varying the crop model, the climate model and the assumption on CO_2 fertilisation, plus combinations of these changes. First, the sensitivity to the crop model and climate model is investigated, and then the effect of CO_2 fertilisation is overlaid to assess the full range of uncertainty for these alternatives.

Figure 2.14 shows projections for four different scenarios, combining different climate and crop models with each other, all without the CO_2 fertilisation effect. At the global level, the central projection is rather close to the results using the alternative IPSL climate model and the default DSSAT crop model, which results in slightly larger negative impacts. Global agricultural damages are projected to be lower if the alternative crop model LPJmL is used, especially in combination with the alternative climate model (IPSL). But regional climate patterns also differ across scenarios, and this ordering does not hold for all regions. All four models project global damages by 2060 in the range of 0.6% to 0.8% of global GDP.

Figure 2.14. **Regional agricultural damages for alternative scenarios, no CO_2 fertilisation**

Percentage change in GDP in 2060 w.r.t. no-damage baseline

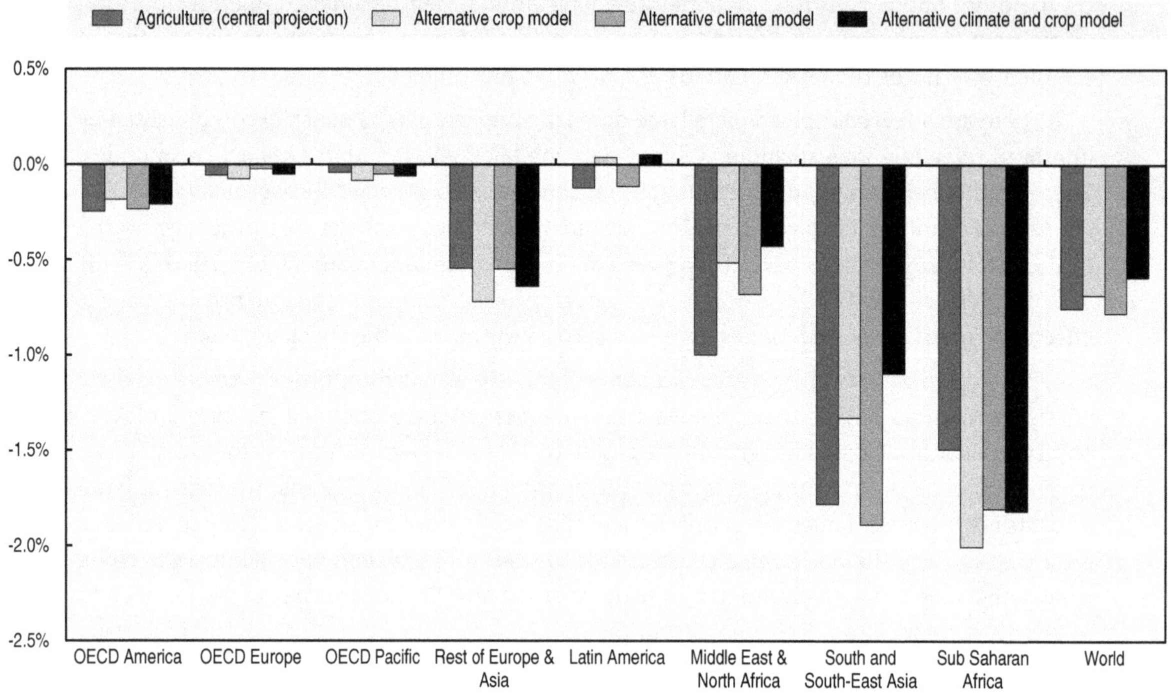

Source: ENV-Linkages model.

StatLink http://dx.doi.org/10.1787/888933276041

Despite some significant differences in the regional projections for these different models (not least in regional projections of water stress), the influence on GDP losses at the level of the 8 macro regions is fairly limited. This reflects the fact that, through trade, the agricultural system operates world-wide, so that to some extent yield losses in one region can be compensated by increased production in other regions. While this may affect specific regions substantially, the consequences for the global food market are much more robust. Nonetheless, the observed differences are strong enough to suggest that more detailed multi-model assessments such as undertaken in AgMIP can provide more robust insights into the consequences of climate change.

As mentioned in Chapter 1, these projections exclude an effect of higher carbon concentrations on crop growth (the CO_2 fertilisation effect, for which the basic idea is that increased concentrations of CO_2 in the atmosphere boost the size and dry weight of crops as plants use CO_2 during photosynthesis). While the data from Nelson et al. (2014) do not provide projections of this effect, the underlying crop model information as reported in Rosenzweig et al. (2013) and synthesised in the Geoshare tool[7] can be used to assess the sensitivity of the economic analysis to this effect. For the default combination of the DSSAT cropmodel and HadGEM climate model, projections on the CO_2 fertilisation effect are available for rice, wheat, maize and soybeans. A wider set of crops, including also rapeseed, millet, sugarcane and sugar beets, is covered by the combination of the LPJmL cropmodel and HadGEM climate model. CO_2 fertilisation effects are not available for fruits and vegetables (including potatoes), and for the plant fibres sector (which includes cotton). For the alternative climate model IPSL, CO_2 fertilisation effects are only provided for non-irrigated lands. For missing data, the assumption is made that there is no CO_2 fertilisation effect, although the literature suggests that for some crops not included here (not least potatoes) the effect may be quite strong (Leakey, 2009).

As an illustration of the projection data that is used in the simulations, Table 2.4 presents the yield shocks for 2050 for the Sub-Saharan Africa region (excl. South Africa). The table shows that the impacts of climate change on crop yields varies widely between crops, and the model choice can even change the sign of the yield effect, as shown in the table for sugarcane/sugar beets. The table also illustrates that the effect of CO_2 fertilisation is quite strong and positive and can limit some of the major negative consequences in agriculture.

Table 2.4. **Impacts of climate change on yields in Sub-Saharan Africa**

Percentage change in yields in 2050 relative to current climate

	Central projection		Alternative crop model		Alternative climate model		Alternative climate and crop model	
	w/o CO_2 effect	w/CO_2 effect	w/o CO_2 effect	w/CO_2 effect	w/o CO_2 effect	w/CO_2 effect	w/o CO_2 effect	w/CO_2 effect
Rice	-23	15	-23	7	-18	-4	-15	-1
Wheat	-18	-4	-21	-5	-29	-12	-23	-5
Other grains	-12	-9	-16	-12	-21	-15	-20	-13
Fruits and veg.	-25	n.a.[1]	-28	n.a.[1]	-22	n.a.[1]	-16	n.a.[1]
Sugarcane and beet	14	n.a.[1]	12	13	-21	n.a.[1]	-20	n.a.[1]
Oilseeds	-31	7	-34	1	-23	8	-19	12
Plant fibres	-31	n.a.[1]	-31	n.a.[1]	-24	n.a.[1]	-19	n.a.[1]

1. Indicates that no projections are available for the CO_2 fertilisation effects for these crops and thus the yield changes without CO_2 fertilisation effect were used in the simulations.

Source: ENV-Linkages model based on AgMIP projections (Von Lampe et al., 2014).

StatLink http://dx.doi.org/10.1787/888933276249

Panel A of Figure 2.15 shows the time profile of global agricultural damages (simulated in isolation, i.e. without the other climate impacts) for the 8 different combinations of two crop models, two climate models, and two choices for CO_2 fertilisation. With the climate and crop models from the central projection, the CO_2 fertilisation effect amounts to 0.2 per cent-points of GDP by 2060, i.e. agricultural damages are a little less than 0.6% of GDP rather than a little less than 0.8%. The simulations with the alternative crop and climate models provide similar gains from CO_2 fertilisation (between 0.2 and 0.3 per cent-points, respectively). The reduced agricultural damages also translate into lower damages from all impacts together, as shown in Panel B. The likely uncertainty range on the ECS shifts from 1.0%-3.3% to 0.9%-3.0%. The percentage-point changes correspond roughly to those of the agricultural damages, indicating that there are limited interaction effects with the economic consequences of the other impact categories.

Finally, Figure 2.16 highlights the regional differences: for some regions, especially OECD Europe and OECD Pacific, the range of the projections of the four model combinations under scenarios of CO_2 fertilisation and no CO_2 fertilization is very small, with minor impacts projected in all scenarios. For other regions, the range is much wider. Relatively large gains from CO_2 fertilisation are projected for the African regions; adding CO_2 fertilisation to the central projection reduces agricultural damages in Sub-Saharan Africa from 1.5% to 1% of GDP. For Latin America, CO_2 fertilisation is also beneficial for crop yields, but a perverse effect occurs in all scenarios: the lower crop yield losses in other regions reduce the trade gains that occur in Brazil, and hence negatively affect Brazil's GDP.

Figure 2.15. **Global damages for alternative agricultural impact scenarios (including CO_2 fertilisation)**

Percentage change in GDP w.r.t. no-damage baseline

A. Agricultural damages for alternative scenarios

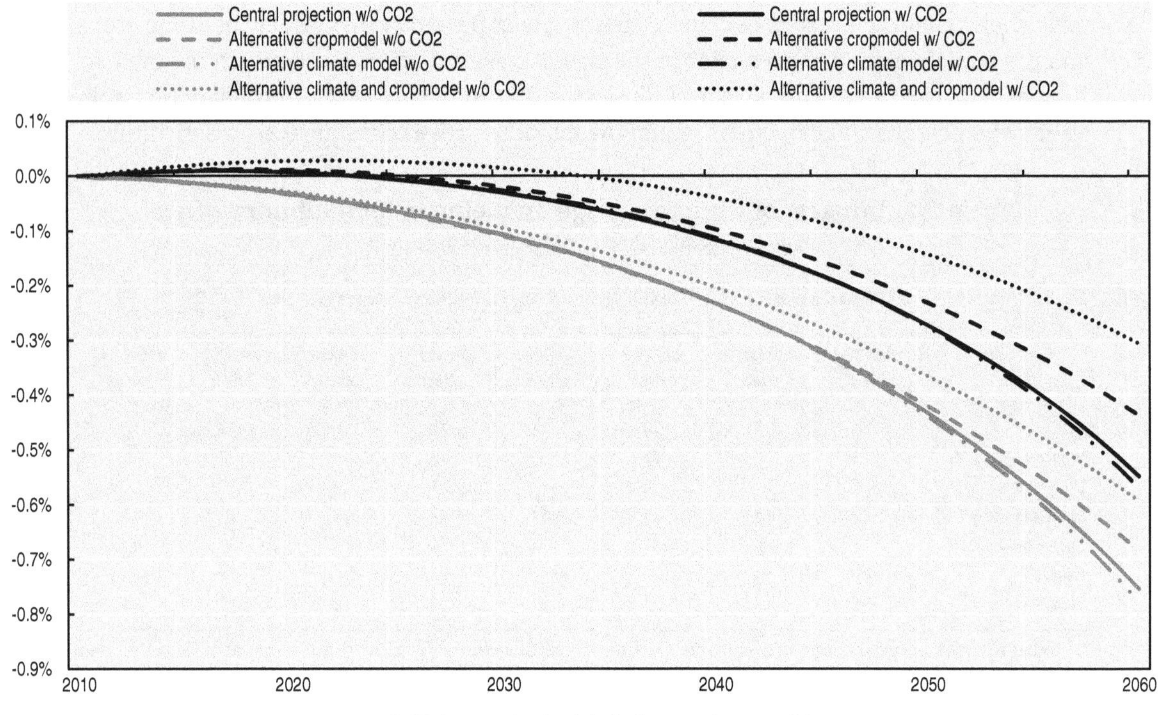

B. All damages for the default climate and crop model

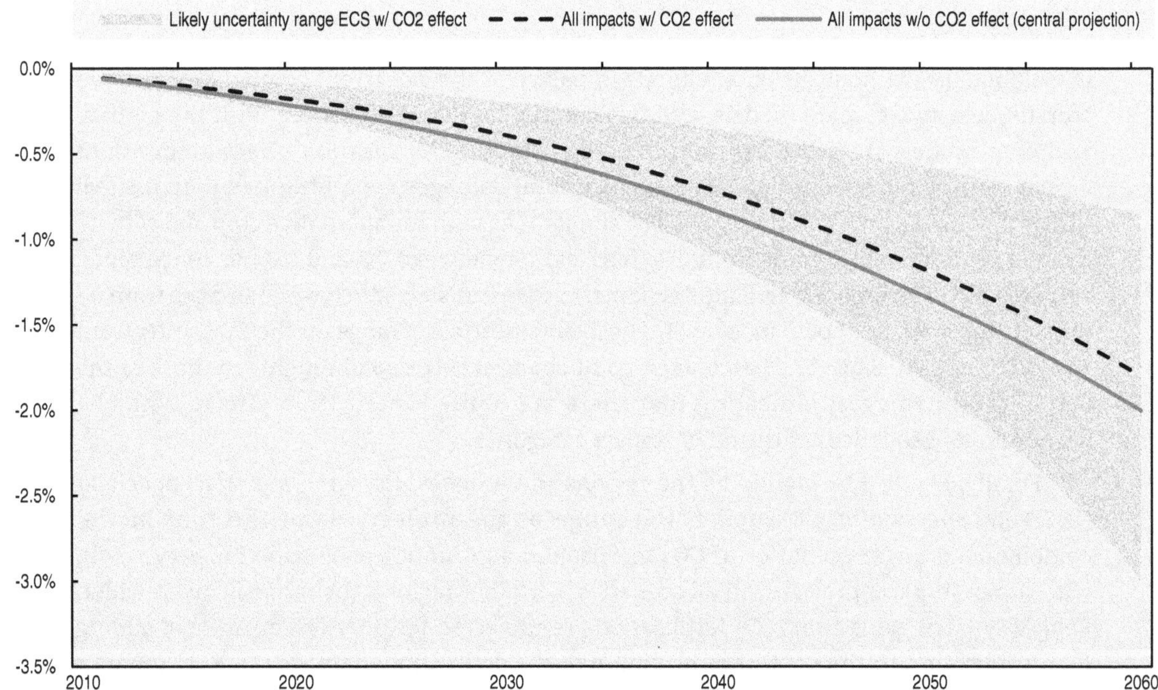

Source: ENV-Linkages model.

StatLink http://dx.doi.org/10.1787/888933276057

In the central projection, the international trade gains outweigh the negative impacts on domestic crop yields (cf. Figure 2.13); in the scenarios that include the CO_2 fertilisation effect, these international trade gains disappear, leading to an overall net negative effect of agricultural impacts on GDP in Latin America.

Figure 2.16. **Range of regional agricultural damages for alternative scenarios (including CO_2 fertilisation)**

Percentage change in GDP in 2060 w.r.t. no-damage baseline

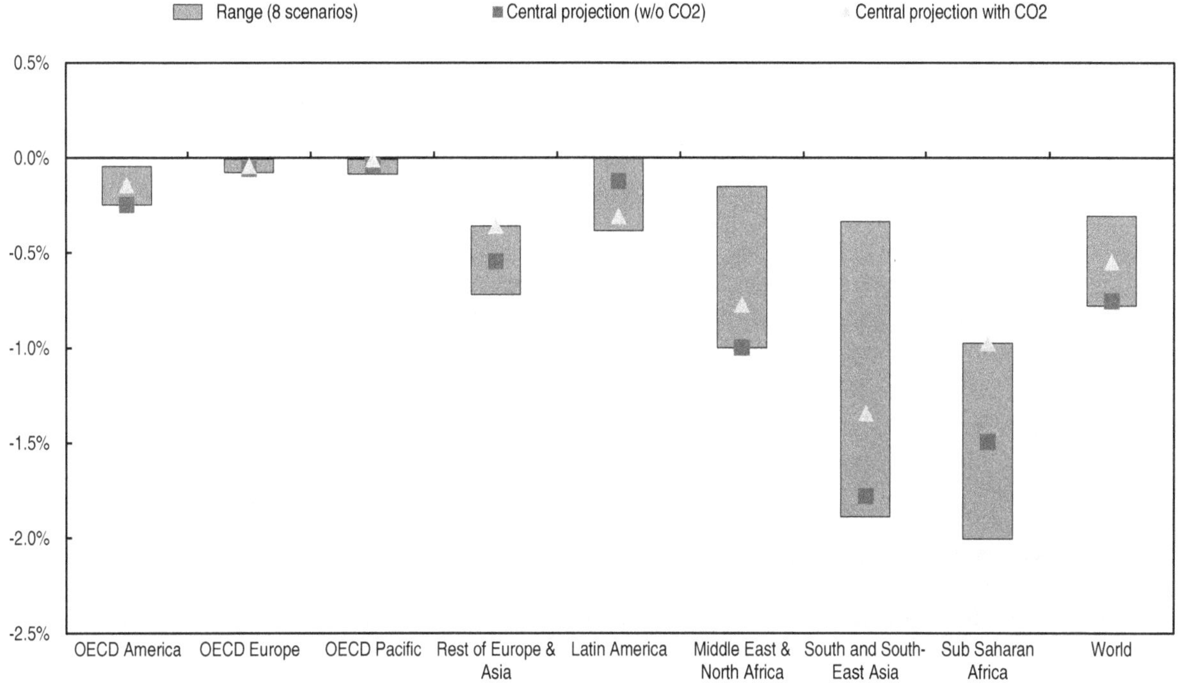

Source: ENV-Linkages model.

StatLink http://dx.doi.org/10.1787/888933276062

Figure 2.16 also demonstrates that the central projection should not be confused with a median projection across scenarios. For OECD America and the Middle East and North Africa, the central projection is actually at the lower edge of the range across scenarios, and the global results are also very close to the lower edge (as can also be seen in Figure 2.15).

Notes

1. More specifically, any policy that is not yet fully implemented, or that still requires an effort to be reached, is not included in the baseline. This assumption is only to provide a reference point for the assessments of the costs of inaction and the benefits of policy action, and does not reflect a view on the state of current climate policies.
2. Alternative population projections are available for the SSP scenarios (Lutz and KC, 2015); for example, in the medium SSP2 scenario, there is a stronger effect of female education on fertility than assumed here, leading to lower population levels later in the century. Using different population projections may substantially affect the numerical analysis in this chapter.
3. An empirical literature is starting to emerge pointing to already occurring climate impacts (Dell et al., 2009, 2013). Although this literature cannot be properly reflected in the long-term projections presented here, the modelling simulations do show small feedback effects on economic growth in the current decade.

4. This is just a crude approximation of the number of people affected by climate change, as many people in countries where overall impacts are positive are negatively affected, either directly by health impacts or indirectly through changes in the domestic economy. Similarly, there will be people in all regions that may benefit from the climate changes or are largely unaffected.
5. The growth rate effects of climate change are further explored in Section 3.1.3.
6. For a robust evaluation of the full uncertainty of climate change impacts on agriculture, a wider range of different models and assumptions should be used.
7. See: *https://mygeohub.org/groups/geoshare.*

References

Akpinar-Ferrand, E. and A. Singh (2010), "Modeling increased demand of energy for air conditioners and consequent CO_2 emissions to minimize health risks due to climate change in India", *Environmental Science and Policy*, Vol. 13(8), pp. 702-712.

Anthoff, D. and R. Tol (2013), "The uncertainty about the social cost of carbon: A decomposition analysis using FUND", *Climatic Change* 117(3), pp. 515-530.

Bosello, F. and R. Parrado (2014), "Climate Change Impacts and Market Driven Adaptation: The Costs of Inaction Including Market Rigidities", *FEEM Working Paper*, 64.2014.

Bosello, F., F. Eboli and R. Pierfederici (2012), "Assessing the Economic Impacts of Climate Change. An Updated CGE Point of View", *FEEM Working Paper,* No. 2.

Braconier, H., G. Nicoletti and B. Westmore (2014), "Policy challenges for the next 50 years", *OECD Economic Policy Paper*, No. 9, July 2014.

Chateau, J., C. Rebolledo and R. Dellink (2011), "An Economic Projection to 2050: The OECD 'ENV-Linkages' Model Baseline", *OECD Environment Working Papers*, No. 41, OECD Publishing, Paris, *http://dx.doi.org/10.1787/5kg0ndkjvfhf-en.*

Ciscar. J.C. et al. (2014), "Climate Impacts in Europe. The JRC PESETA II Project", *JRC Scientific and Policy Reports*, No. EUR 26586EN, Luxembourg: Publications Office of the European Union.

Criqui, P. (2001), "POLES: Prospective outlook on long-term energy systems", *Institut d'Économie et de Politique de l'Énergie*, No. 9.

Criqui, P., S. Mima and P. Menanteau (2009), "Trajectories of new energy technologies in carbon constraint cases with the POLES Model", *IOP Conference Series: Earth and Environmental Science*, No. 6 (2009) 212004.

Dell, M., B.F. Jones and B.A. Olken (2013), "What do we learn from the weather? The new climate-economy literature", *NBER (National Bureau of Economic Research) Working Paper Series*, No. 19578, NBER, Cambridge, Massachusetts.

Dell, M., B.F. Jones and B.A. Olken (2009), "Temperature and Income: Reconciling New Cross-Sectional and Panel Estimates", *American Economic Review*, Vol. 99(2), pp. 198-204.

Dellink, R. et al. (2014), "Consequences of Climate Change Damages for Economic Growth: A Dynamic Quantitative Assessment", *OECD Economics Department Working Papers*, No. 1135, OECD Publishing, Paris, *http://dx.doi.org/10.1787/5jz2bxb8kmf3-en.*

Dietz, S., C. Gollier and L. Kessler (2015), "The climate beta", Grantham Research Institute on Climate Change and the *Environment Working Paper* 190, London.

Eboli, F., R. Parrado and R. Roson (2010), "Climate-change feedback on economic growth: Explorations with a dynamic general equilibrium model", *Environment and Development Economics*, Vol. 15, pp. 515-533.

Electric Power Research Institute (EPRI) (2014), "Understanding the social cost of carbon: a technical assessment", *EPRI Report* #3002004657, Palo Alto.

Gollier, C. (2012), *Pricing the planet's future: The economics of discounting in an uncertain world*, Princeton University Press.

Gollier, C. (2007), "Comment intégrer le risque dans le calcul économique?", *Revue d'économie politique*, Vol. 117(2), pp. 209-223.

Heal, G. and A. Millner (2013), "Uncertainty and decision in climate change economics", *NBER Working Paper* No. 18929; also published as Grantham Research Institute on Climate Change and the *Environment Working Paper* 108, London.

Hertel, T.W., M.B. Burke and D.B. Lobell (2010), "The poverty implications of climate-induced crop yield changes by 2030", *Global Environmental Change*, Vol. 20, pp. 577-585.

Hinkel, J. et al. (2014), "Coastal flood damage and adaptation costs under 21st century sea-level rise", *PNAS*, Vol. 111(9), pp. 3292-97, doi: *10.1073/pnas.1222469111*.

IEA (2014), *World Energy Outlook 2014*, IEA, Paris, *http://dx.doi.org/10.1787/weo-2014-en*.

IEA (2013), *Redrawing the energy climate map*, IEA, Paris.

IMF (2014), *World Economic Outlook*, Washington, DC.

IPCC (2014a), *Climate Change 2014: Impacts, Adaptation, and Vulnerability. Part A: Global and Sectoral Aspects*, Contribution of Working Group II to the Fifth Assessment Report of the Intergovernmental Panel on Climate Change [Field, C.B., V.R. Barros, D.J. Dokken, K.J. Mach, M.D. Mastrandrea, T.E. Bilir, M. Chatterjee, K.L. Ebi, Y.O. Estrada, R.C. Genova, B. Girma, E.S. Kissel, A.N. Levy, S. MacCracken, P.R. Mastrandrea, and L.L.White (eds.)]. Cambridge University Press, Cambridge, United Kingdom and New York, NY, USA, 1132 pp.

IPCC (2014b): *Climate Change 2014: Mitigation of Climate Change*, Contribution of Working Group III to the Fifth Assessment Report of the Intergovernmental Panel on Climate Change [Edenhofer, O., R. Pichs-Madruga, Y. Sokona, E. Farahani, S. Kadner, K. Seyboth, A. Adler, I. Baum, S. Brunner, P. Eickemeier, B. Kriemann, J. Savolainen, S. Schlömer, C. von Stechow, T. Zwickel and J.C. Minx (eds.)]. Cambridge University Press, Cambridge, United Kingdom and New York, NY, USA.

IPCC (2013), *Climate Change 2013: The Physical Science Basis*, Contribution of Working Group I to the Fifth Assessment Report of the Intergovernmental Panel on Climate Change [Stocker, T.F., D. Qin, G.-K. Plattner, M. Tignor, S.K. Allen, J. Boschung, A. Nauels, Y. Xia, V. Bex and P.M. Midgley (eds.)]. Cambridge University Press, Cambridge, United Kingdom and New York, NY, USA, 1535 pp.

Leakey, A.D.B. (2009), "Rising atmospheric carbon dioxide concentration and the future of C4 crops for food and fuel", *Proceedings of the Royal Society B*, Vol. 276(1666), pp. 2333-43.

Meinshausen, M., S.C.B. Raper and T.M.L. Wigley (2011), "Emulating coupled atmosphere-ocean and carbon cycle models with a simpler model, MAGICC6: Part I – Model Description and Calibration", *Atmospheric Chemistry and Physics*, Vol. 11, pp. 1417-56.

Mima S., P. Criqui and P. Watkiss (2011), "The Impacts and Economic Costs of Climate Change on Energy in Europe. Summary of Results from the EC RTD ClimateCost Project."; in: P. Watkiss (ed.), *The ClimateCost Project. Final Report. Volume 1: Europe*, Published by the Stockholm Environment Institute, Sweden, *www.climatecost.eu*.

Nelson, G.C. et al. (2014), "Climate change effects on agriculture: Economic responses to biophysical shocks", *Proceedings of the National Academy of Sciences*, Vol. 111(9), 3274-79.

Nordhaus, W.D. (2013), *The climate casino: Risk, uncertainty, and economics for a warming world*, Yale University Press, New Haven.

Nordhaus, W.D. (2011), *Estimates of the social cost of carbon: Background and results from the RICE-2011 model*, Conn. Cowles Foundation for Research in Economics, Yale University, New Haven.

Nordhaus, W.D. (2007), *A question of balance*, Yale University Press, New Haven, United States.

OECD (2014), *OECD Economic Outlook*, Vol. 2014/1, OECD Publishing, Paris, *http://dx.doi.org/10.1787/eco_outlook-v2014-1-en*.

Otto, A. et al. (2013), "Climate system properties determining the social cost of carbon", *Environmental Research Letters*, Vol. 8(2), doi: *10.1088/1748-9326/8/2/024032*.

Rogelj, J., M. Meinshausen and R. Knutti (2012), "Global warming under old and new scenarios using IPCC climate sensitivity range estimates", *Nature Climate Change*, Vol. 2, pp. 248-253.

Rosenzweig, C. et al. (2013), "Assessing agricultural risks of climate change in the 21st century in a global gridded crop model intercomparison", *Proceedings of the National Academy of Sciences (PNAS)*, Vol. 111(9), pp. 3268-73.

Roson, R. and D. Van der Mensbrugghe (2012), "Climate change and economic growth: impacts and interactions", *International Journal of Sustainable Economy*, Vol. 4, pp. 270-285.

Schellnhuber, H.J., K. Frieler and P. Kabat (2013), "The elephant, the blind, and the ISI-MIP", *Proceedings of the National Academy of Sciences (PNAS)*, Vol. 111(9), pp. 3225-27.

Tol, R.S.J. (2012), "On the uncertainty about the total economic impact of climate change", *Environmental and Resource Economics* 53, pp. 97-116.

Tol, R.S.J. (2002), "New Estimates of the Damage Costs of Climate Change, Part I: Benchmark Estimates", *Environmental and Resource Economics*, Vol. 21 (1), pp. 47-73.

United Nations (2013), *World Population Prospects: The 2012 Revision*, UN Department of Economic and Social Affairs.

Van Vuuren, D.P., E. Kriegler, B.C. O'Neill, K.L. Ebi, K. Riahi, T.R. Carter, J. Edmonds, S. Hallegatte, T. Kram, R. Mathur, H. Winkler (2014), "A new scenario framework for climate change research: Scenario matrix architecture", *Climatic Change*, Vol. 122, pp. 373-386.

Von Lampe, M. et al. (2014), "Why do global long-term scenarios for agriculture differ? An overview of the AgMIP Global Economic Model Intercomparison", *Agricultural Economics*, Vol. 45(1), pp. 3-20.

Watkiss, P. and A. Hunt (2012), "Projection of economic impacts of climate change in sectors of Europe based on bottom up analysis: Human health", *Climatic Change*, Vol. 112, pp. 101-126.

Wilby, R.L. and S. Dessai (2010), "Robust adaptation to climate change", *Weather*, Vol. 65(7), pp. 180-185, doi: *0.1002/wea.543*.

Chapter 3

The bigger picture of climate change*

Without aiming to be complete, this chapter provides insights into the main consequences of inaction that are not assessed in Chapter 2. The chapter starts with an analysis using the integrated assessment model AD-DICE to shed light on the long-term consequences of climate change. It then presents a number of important examples of market and non-market consequences of climate change that are not included in the analysis with the ENV-Linkages model. Two prominent examples of these are the projected urban damages from river floods and the welfare costs from premature deaths due to heat stress.

* This document and any map included herein are without prejudice to the status of or sovereignty over any territory, to the delimitation of international frontiers and boundaries and to the name of any territory, city or area.

The numerical results presented in Chapter 2 reveal the scale of the costs of the selected impacts of climate change to the economy in the absence of new policies to respond to the risks of climate change. However, merely considering the effects on Gross Domestic Product (GDP) produced by the ENV-Linkages model will fall short of representing the entirety of the economic and social consequences of higher global average temperature, rising sea level, and other climatic drivers. At least two further aspects need to be taken into account for a more appropriate evaluation of the costs of climate change: i) the long-term consequences of climate impacts on the economy after 2060, including those that stem from emissions before 2060, and ii) the major consequences of climate change that cannot be captured in the general equilibrium model projections of GDP changes. The inclusion of these factors into the assessment of the costs of inaction is also fundamental to appropriately assess the benefits of climate change policy action.

3.1. Costs of inaction beyond 2060

Greenhouse gases emitted by 2060 will affect the climate and the economy until 2060, but will also have important consequences in the decades and even centuries that will follow. Solely projecting GDP impacts before 2060 will therefore underestimate the net present value of GDP impacts of climate change as the long-term impacts – due to the inertia of the climate system – are ignored. The AD-DICE2013 model (De Bruin, 2014) is used to study the long-term consequences of climate change through stylised simulations.

In this section, first the central projection scenario is extended to 2100, to highlight how climate damages are projected to increase over time: this is referred to as the *Full damages* scenario.[1] In contrast to the analyses of Nordhaus using the DICE model (e.g. Nordhaus, 2012), these projections adopt the discount factors that are recommended by the UK Treasury (2003), and limit adaptation only to market-induced adaptation actions (see Section 4.2 and Annex I for more details on this). The latter difference implies that damages as projected in this section by AD-DICE are higher than those projected by the standard specification of the DICE model; Annex I provides an overview of where the differences between AD-DICE and the original DICE model come from. Furthermore, the level of damages in AD-DICE until 2060 has been calibrated to the central projection of the ENV-Linkages model, as described in Chapter 2. Consequently, these long-term projections still capture the same subset of all impacts, albeit in a more stylised way.

Second, the *Committed by 2060* scenario projects how 50 years of inaction will affect the future estimated GDP impacts of climate change. The lack of policy action to reduce emissions prior to 2060 will affect GDP levels over time due to climate change set into motion in the period before 2060. Damages associated with inaction before 2060 will continue after 2060. In the Committed by 2060 scenario this is modelled by setting emissions at the business as usual level until 2060, after which economic production is decoupled from emissions (and emissions are set to zero). This hypothetical scenario can shed light on the irreversible, unavoidable level of climate change that the world is committing to by 2060 if no further adaptation or mitigation policies are adopted until then.

Third, an alternative damage specification is used that puts larger weights on the tail risks of climate damages, i.e. the possibility that high levels of climate change lead to much more significant economic consequences than assumed in the gradual damage function that is normally employed in AD-DICE (cf. Weitzman, 2012). As long-term consequences of climate change that go beyond 4 degrees global average temperature increase are poorly understood, the alternative specification sheds light on the risks of underestimating impacts at these unprecedented levels of climate change.

Finally, a stylised scenario analysis is carried out to shed light on the important distinction whether climate damages affect only the level of economic output, or primarily the economic growth rate. In the latter case, the long-term consequences may be much more severe, as impacts lead to lower future economic growth.

3.1.1. *Long-term consequences of delayed action*

By construction, the central projection in AD-DICE leads to global GDP losses by 2060 that are in line with those projected in ENV-Linkages (and that represent the same subset of impacts, as outlined in Chapter 1). If emissions continue to rise after 2060, the negative effect on GDP continues and central projections of GDP losses become 5.8% of GDP by the end of the 21st century according to the AD-DICE model (Figure 2.15). As in the ENV-Linkages calculations, these projections are explicitly linked to uncertainty on the equilibrium climate sensitivity (ECS) parameter, but are subject to other uncertainties as well (cf. Box 2.1). By the end of the 21st century, GDP losses for the likely ECS range of 1.5°C and 4.5°C are projected to be between 2% and 10% (for the wider range for ECS values between 1°C and 6°C the range in GDP losses is 1% to 15%). The associated global average temperature increase by 2100 as projected by AD-DICE is between 2.4 and 5.5°C for the likely range. These global damages in AD-DICE are larger than the Business-as-Usual projections presented by the various versions of the DICE model (Nordhaus, 2007, 2012), and also than the results for AD-RICE as presented in Dellink et al. (2014). While the recalibration of the AD-DICE model to match the CIRCLE baseline may have a minor effect on the results compared to DICE (De Bruin, 2015), by far the largest difference between AD-DICE and DICE/RICE stems from the assumption in the central projection in AD-DICE that not all adaptation options are implemented without new climate policies (cf. Section 4.2.1 and Appendix I). Dellink et al. (2014) used a slightly older version of AD-RICE that excludes sea level rise damages and hence leads to smaller GDP impacts, and adopts full adaptation as the central projection.

The *Committed by 2060* scenario by construct sets emissions at their business as usual levels until 2060, after which emissions are set to zero. Comparing the *Full damages* scenario and the *Committed by 2060* scenario, Figure 3.1 shows that a substantial part of these impacts are already locked-in by the emissions occurring until 2060. Especially sea-level rise damages respond only very slowly to a change in the emission pulse. Even if net greenhouse gas (GHG) emissions dropped to zero after 2060, climate damages and consequent effects on GDP (of around 2% by 2060) would continue to increase for at least a century more due to the inertia in the climate system (stabilising at 3% later in the century).[2] This is in line with findings by the Intergovernmental Panel on Climate Change (IPCC, 2013), which stresses that "surface temperatures will remain approximately constant at elevated levels for many centuries after a complete cessation of net anthropogenic CO_2 emissions" (IPCC, 2013, p. 26). Even if climate sensitivity in equilibrium is low (i.e. 1.5°C), annual GDP losses of at least 1% are committed to for over a century after 2060.

Figure 3.1. **Climate change damages from selected climate change impacts in the very long run, central projection**

Percentage change w.r.t. no-damage baseline

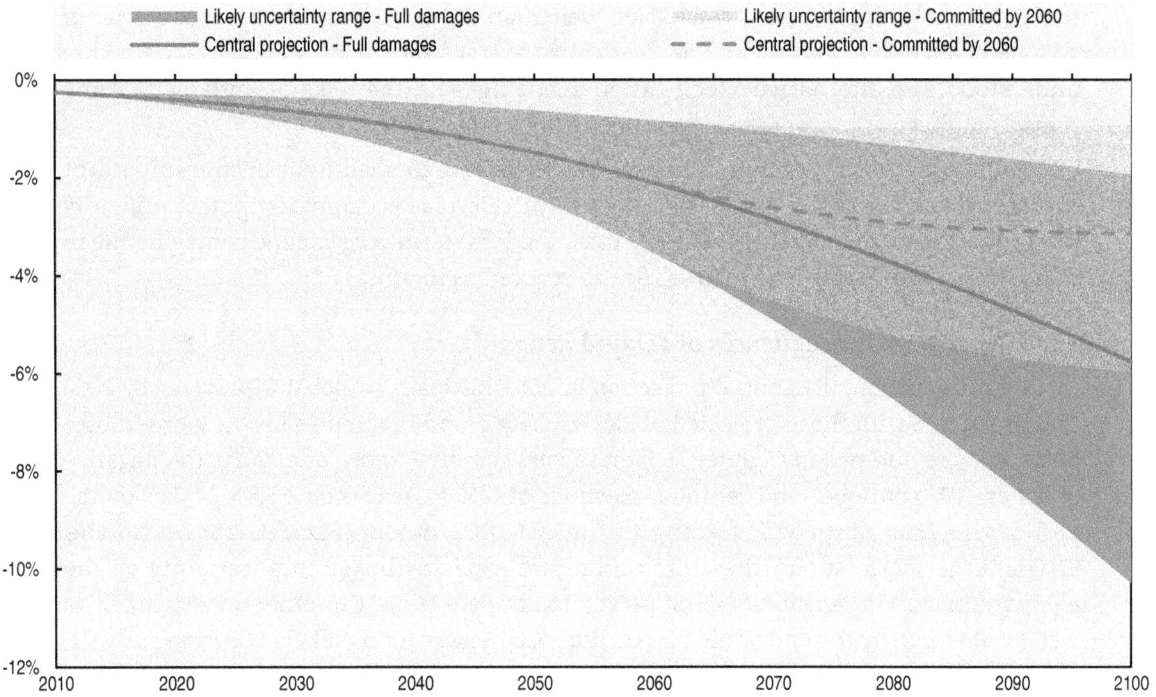

Source: AD-DICE model.

StatLink http://dx.doi.org/10.1787/888933276073

In the case of a high climate sensitivity (equal to 4.5°C or 6°C temperature increase), this annual loss rises to 6% and more than 9%, respectively, by 2100. This insight also holds for climate impacts occurring before 2060: effectively any emission, whether now or in the future, triggers a series of effects and leads to an increase in climate damages for at least a century. Thus, there are damages that are already committed to now due to historical emissions; in the AD-DICE model, these gradually increase to around 0.6% of GDP (for the central ECS estimate), although the model is not fine-grained enough (and not intended to be) to assess current damage levels accurately.

3.1.2. *An alternative damage function with stronger non-linearity*

Long-term GDP impacts crucially depend on the shape of the damage function. However, as the IPCC Working Group III (2014b) stresses, the higher temperature increases are, the less robust the damage projections. AD-DICE uses the typical quadratic damage function as proposed by Nordhaus, but Weitzman (2009, 2012, 2013) has argued that this underestimates small-likelihood, high-impact possibilities (the so-called "fat tail"). An alternative damage function is calibrated for AD-DICE following Weitzman (2012) which exhibits much stronger non-linearity. Temperature increases up to 2°C lead to similar impact levels as in the standard specification of the AD-DICE model, but large temperature increases lead to much more dramatic reductions in GDP (by including a higher power term in the damage response to temperature increases, as provided by Weitzman, 2012). As shown in Figure 3.2, the long-term consequences of this alternative specification are dramatic in the central scenario without policy action, where GDP impacts go into double

Figure 3.2. **Climate change damages from selected climate change impacts in the very long run, alternative damage function**

Percentage change w.r.t. no-damage baseline

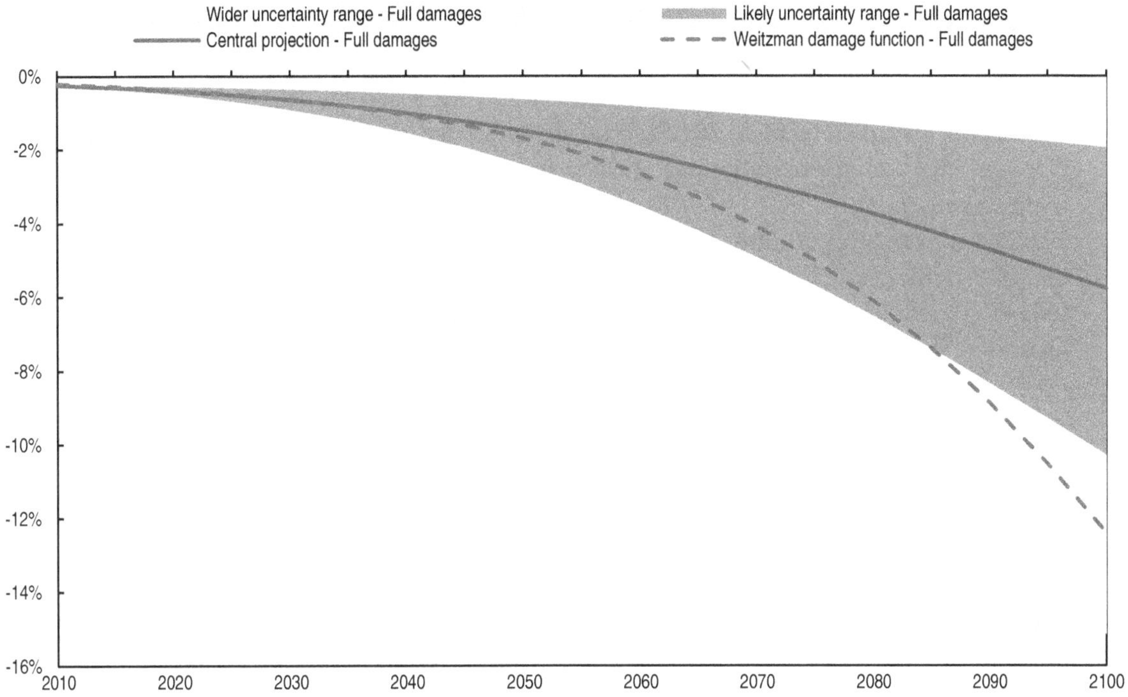

Source: AD-DICE model.

StatLink http://dx.doi.org/10.1787/888933276087

digits before the end of the 21st century. Note that in the *Committed by 2060* scenario with the Weitzman damage function, the losses in the later decades of the 21st century are also markedly higher than in the base specification with the original damage function of the AD-DICE model (5% versus 3%; not shown in Figure 3.2).

3.1.3. *Climate change damages directly affecting economic growth rates*

There is an emerging literature that suggests that IAMs should implement damages at least partially as an impact directly on the growth rate of the economy, rather than as an impact on the level of GDP as in DICE (Nordhaus, 2012), AD-DICE and similar models (see Box 3.1). The basic idea is that unabated climate change may disrupt the "engines of growth" (such as capital accumulation and technological progress). For instance, Dell et al. (2012) find empirical evidence that temperature shocks affect economic growth rates in developing countries. Felbermayr and Gröschl (2014) similarly find persistent effects for natural disasters and Hsiang and Jina (2014) for windstorms in particular. This is especially relevant for developing countries, which require a rapid increase in capital stocks and a fast development of technology in order to catch up with more developed countries in terms of income levels. Such growth effects can stem from a change in the technological growth rate of the economy or from destruction of capital stocks. The logic of assuming that damages affect the technological growth rate (specifically, total factor productivity, TFP) is that current technologies are geared towards current climate conditions, and may not work as well in future climate conditions (Dietz and Stern, 2015). Similarly, the rationale for the effect on the capital stock is that at least some damages come in the form of a destruction of land, buildings, etc. Sue Wing and Lanzi (2014) and Chapters 1 and 2 also highlight the rationale for capital stock damages.

AD-DICE is uniquely placed to shed further light on all these elements, and tease out to what extent placing damages directly on growth is important for policy evaluation. Hence, two additional simulations are carried out in which part of the residual damages (i.e. excluding adaptation costs) are re-allocated to the drivers of growth. In the first simulation, 30% of all residual damages are allocated to capital stock. The partition at 30% is in line with both the assumptions by Dietz and Stern (2015), and with the observation in the previous chapter that by 2060, roughly half of the GDP losses are induced by a slowdown of capital accumulation. Secondly, a variant is simulated in which 30% of residual damages (excluding adaptation costs) affect the growth rate of TFP, again following Dietz and Stern (2015). In both cases, the remaining 70% of residual damages, plus all adaptation costs, are allocated to GDP.

Box 3.1. **An emerging literature on climate damages affecting economic growth**

Pindyck (2012), Stern (2013) and Dietz and Stern (2015) take the critical assumption that the climate impacts as applied in DICE (and AD-DICE) to the level of GDP should at least partially be applied to the growth rate of TFP. However, technically this would only be valid when the impacts are estimated independently of GDP. If the impacts to be reproduced in the integrated assessment model are presented in the form of a stream of annual macroeconomic losses resulting from a stream of impacts, then these annual flows should be recalibrated when applied to growth rates to mimic the (exogenous) stream of impacts as percentage of GDP. Not recalibrating the input parameters in this case would lead to double-counting, as future damages from current impacts are already included in the future stream of losses. This recalibration implies that the formulation with damages allocated to growth rather than to the level of income would have no net effect on future projections, but merely entails a change in the functional form of the impacts. With stylised aggregated damage functions such as the one in DICE and AD-DICE, it is impossible to identify the extent to which damages need to be recalibrated or not when they are applied to growth rates.

Stern (2013) and Dietz and Stern (2015) also investigate the option that climate damages affect the stock of capital. The detailed impact assessment in ENV-Linkages incorporates both effects: some of the impacts affect growth rates (e.g. sea level rise and hurricane damages to capital stocks), while others affect levels of output (e.g. agricultural crop yield impacts).

Moore and Diaz (2015) go one step further and calibrate the growth impact of climate change to the empirical results that historically, a higher regional temperature has been negatively correlated with economic growth rates, at least for poor countries (Dell et al., 2012). Moore and Diaz (2015) take these estimates and apply it to future economic growth projections. There are several reasons for concern with this approach. First of all, most credible baseline projections, including their own, show that future growth rates are expected to be higher in currently poor countries than in richer countries (in technical terms, damages and income are negatively correlated). This is the conditional convergence assumption, which features e.g. in OECD's economic projections and also underlies the CIRCLE baseline (see Chapter 2). Damages may reduce the speed of convergence (by affecting growth rates more severely in developing countries), or hamper development in

Box 3.1. **An emerging literature on climate damages affecting economic growth** *(cont.)*

other ways such that convergence is slower than without climate change. But climate change will have to have very strong consequences for growth to prevent all convergence. And if the convergence effect dominates the climate damage effect, it may well be that in the future economic growth rates are positively correlated with regional temperature levels, rather than negatively. Secondly, compared to historical income levels, almost all countries are projected to be rich by the end of the century (Moore and Diaz also have poor country income levels well above the current global average). As Dell et al. (2012) show that the impact of temperature increase on growth is not significant for higher income countries, this could suggest that growth impacts on currently poor countries also diminishes over time.

The inclusion of adaptation costs in the damage function, such as is usual in the DICE model (but not in AD-DICE), also complicates the picture: the impacts from flooding due to sea level rise come largely in the form of capital destruction, but in a least-cost mix of impacts and adaptation (as adopted in e.g. DICE), in most regions the largest part of the costs are actually investments in sea defense systems. These are investment costs, which come at the expense of consumption and compete with other investments, but do not directly affect the capital stock. In principle, only some of the residual impacts should then be allocated to directly affect the existing capital stock.

The results for these simulations are shown in Figure 3.3. Both alternative specifications lead to an increase in the projected damages over time, and the gap between the different specifications becomes larger over time. But the effects are not dramatic, and despite the implementation of growth effects immediately from 2010, for capital stock damages the differences are hardly visible until the middle of the century. For the capital damages specification, the key factor limiting the damages over time is the increase of the savings rate, to account for the lower capital stock and hence higher marginal productivity of capital. The model automatically finds the optimal trade-off here, in order to maximise the net present value of utility. This adjustment of savings and investment behaviour is a powerful adaptation measure, and largely negates the growth effects of the capital or TFP growth damages. Effectively, households have an instrument they can use to transform growth damages into level damages: by saving more they sacrifice a little bit of consumption in the short term in order to compensate for future capital stock losses and preserve higher growth rates of GDP. The effect of the capital damages on consumption levels (not shown in Figure 3.3) is very similar to the effect on GDP; only a specification with fixed savings rates will produce projections where the effect of capital stock damages are substantially different.

The effect of placing part of the damages on TFP has a larger effect on global GDP levels, of the same order of magnitude as a high ECS or the Weitzman damage function specification. Hence, it does not alter the qualitative insights from the modelling exercise. Moreover, it is conceptually hard to imagine how a large part of climate damages would not affect capital or production, but rather the technological growth rate of economies. In fact, destruction of outdated existing capital may spur technological innovation and potentially increases the efficiency of economies. None of the impact categories included in the detailed quantitative assessment in Chapter 2 is directly related to the TFP growth rate, while some do affect TFP levels.

Figure 3.3. **Climate change damages from selected climate change impacts in the very long run, alternative effects on growth**

Percentage change w.r.t. no-damage baseline

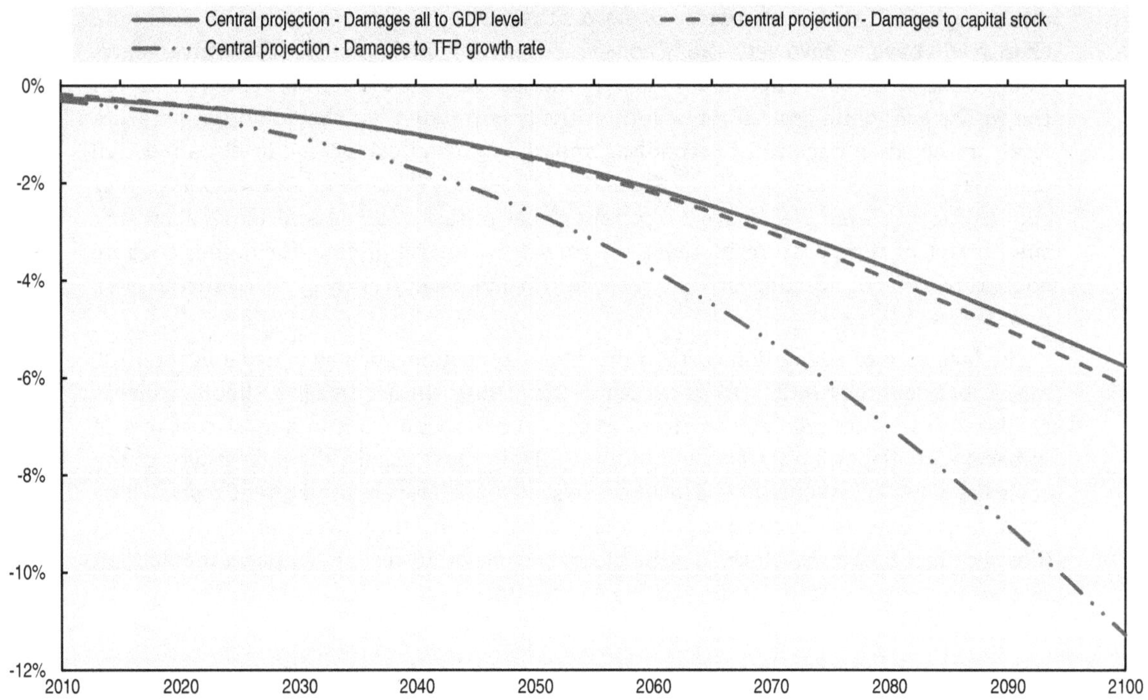

Central projection - Damages all to GDP level
Central projection - Damages to capital stock
Central projection - Damages to TFP growth rate

0%
-2%
-4%
-6%
-8%
-10%
-12%
2010 2020 2030 2040 2050 2060 2070 2080 2090 2100

Source: AD-DICE model.

StatLink http://dx.doi.org/10.1787/888933276090

Both alternative specifications are within the uncertainty range presented for the central projection in Figures 3.1 and 3.2. Thus, whether climate damages affect only levels of GDP or also the growth rate may be theoretically important, but is not of overwhelming importance for applied analysis. While stylised analyses with bold assumptions may provide clear insights into the mechanisms that are at work, this should be complemented with a nuanced approach, by looking at how different climate impacts affect the different drivers of growth, as carried out in Chapter 2 with the ENV-Linkages model, to provide robust insights for actual policy making.

3.2. Other consequences of climate change: Mortality, floods, and tipping points

The analysis with ENV-Linkages aims to take into account the most significant market-based impacts of climate change. Yet, there are several reasons why it cannot be comprehensive. First, for some types of climate impacts, there is insufficient data or knowledge to robustly incorporate them in ENV-Linkages. Secondly, ENV-Linkages is an economic model based on a production-based measurement of economic activity, and thereby has only limited capacity to quantify the consequences of climate change, and especially those not directly related to markets.

Non-market impacts of climate change represent consequences that affect human and non-human activities for which no established economic markets exist (IPCC, 2014a), such as biodiversity and culture, as well as changes in welfare that are not fully captured by GDP, such as the welfare costs from premature deaths or pain and suffering.

The total non-market impacts of climate change are likely to be significant, although there are insufficient comprehensive quantitative analyses to draw robust conclusions. The Stern Report (Stern, 2007), being one of the few exceptions in the literature accounting for such impacts, estimates the non-market damages from climate change to add up to roughly 6% of GDP in a warming scenario of 7.4°C. The latest IPCC Assessment Report does not cite figures on the magnitude of non-market impacts (IPCC, 2014a) due to a lack of availability of robust estimates.

As non-market impacts cannot be appropriately measured with a production-based indicator such as GDP, alternative indicators that have a wider range have emerged as potential alternatives to GDP (Box 3.2).

Box 3.2. **Beyond GDP: The effects of climate change on well-being**

There is ample evidence that GDP is not very good at measuring economic welfare, let alone well-being. Issues such as inequality, health, and environmental quality are just some of the major factors affecting people that are incorporated imperfectly, if at all, in GDP. A thorough overview of the discussion is given in Stiglitz et al. (2010), and the OECD's Better Life Index (OECD, 2011) is one example of interesting new developments in this terrain.

Climate change will have many effects that cannot be appropriately measured with GDP. For instance, premature mortality has negative consequences for life expectancy, and the welfare costs of sick days are more completely measured when including loss of wellbeing than when looking at labour productivity alone. Moreover, the true welfare costs from e.g. loss of cultural heritage can also not be captured by GDP changes. Hence, it makes sense to adopt an approach that does not only measure the consequences of climate change with a narrow economic indicator such as GDP, but employs measures that can express broader welfare consequences. Initiatives aiming to develop such alternative indicators often refer to the expression "Beyond GDP" to describe this kind of work (OECD, 2011; EC, 2009a; 2009b; Costanza et al., 2009; UNDP, 1990, 2014).

While theoretically interesting and relevant, two reasons preclude adopting such an approach in this report. First, the data requirements for a more comprehensive social welfare evaluation of climate impacts are daunting, and in many cases no robust data exist, as also highlighted throughout this chapter. Secondly, GDP still plays a central role in assessing whether societies are prospering, and in order to influence the decision making processes at all levels of government, the focus on GDP allows for a focus of the discussion on impacts and costs rather than on indicators and methodology. Simply put, everyone knows GDP and almost everyone knows at least roughly what's wrong with it.

One possible way to narrow the gap is to broaden the definition of welfare as measured in the economic modelling framework and present results in terms of welfare costs next to GDP losses. For instance, a welfare-cost approximation of the disutility costs from morbidity and mortality could then be brought into the analytical evaluation. This approach is adopted in e.g. Ciscar et al. (2014), building on Mayeres and Van Regemorter (2008). This would also open the possibility to investigate the consequences on wellbeing through risk aversion by calculating a risk premium for the lack of knowledge of what will happen under climate change (Markandya et al., 2015). Such an elaboration is far from straightforward and left for future research.

This section aims to complement the modelling analysis with a selection of examples of relevant other climate change impacts from the literature, including both missing market as well as non-market consequences. For each of the impact categories presented in Chapter 1, one or a few examples will be provided, without attempting to be complete.[3] Where possible, quantitative information based on model projections or direct valuation is provided. This discussion intends to raise readers' awareness of these potential effects and, where relevant, provides useful information for a future integration of some of these impacts into the ENV-Linkages model. While a fully-fledged quantitative analysis of the broader well-being effects of climate impacts is not part of this study, policy makers will need to take a broader view when evaluating specific policy proposals, rather than treat the purely economic consequences presented in this report as the full extent of benefits against which to compare the costs of policies.

3.2.1. Agriculture

The majority of the literature, including the quantitative analysis in this report, relies on indicators related to changes in crop productivity to estimate the costs of climate change to agriculture (IPCC, 2014a). However, climate change will have consequences for various other production and non-production aspects of agriculture, including livestock, pasture- and rangeland, and aquaculture (OECD, 2001; OECD, 2014a).

Livestock is likely to be considerably impacted on by climate change. It is an important part of the agricultural sector and the food supply of both OECD and non-OECD countries. Although the effects of climate change on livestock are much less exhaustively explored than crop production, the largest part of the literature finds negative effects of climate change, not least through heat and water stress, on animal growth, their health, and the commodities they produce, e.g. dairy (IPCC, 2014a). There is, however, a lack of studies with a global coverage of the impacts of climate change on livestock production (IPCC, 2014a). Heat stress, which is projected to increase with climate change (see below), can have significant effects on livestock mortality. Changing precipitation patterns as well as amplified need of cattle and other domesticated animals for water to cope with higher temperatures will likely contribute to this challenge. A study by Wall et al. (2010), for example, shows for the United Kingdom that heat stress induced by climate change can lead to increases in mortality and economic losses from dairy production, amounting to annual losses of about GBP 40 million by the 2080s under a medium to high GHG emissions scenario. In the United States, several states have respectively reported more than 5 000 animals deaths from single heat wave events in the past (USGCRP, 2009). Decreased cold exposure from higher average temperatures could be positive for livestock production, but has not been rigorously explored in the literature and numerical estimates are largely absent. In addition, climate change may increase the incidence of diseases among livestock, especially for ailments transmitted through vectors that are highly dependent on climate conditions (IPCC, 2014a). While experts are highly confident that climate change will spur the spread of diseases, evidence for this relationship is small. Other studies suggest that there will be positive or non-measurable effects of climate change to livestock in some regions. Graux et al. (2011), for instance, cannot identify changes in future dairy yields from climate change in France. Large-scale commercial farmers that rely extensively on cattle may be more sensitive to changes in temperatures than small farms that may more easily switch their production process or the type of animals they breed, as Seo and Mendelsohn (2008) show for Africa.

Figure 3.4. **Climate change impacts on livestock production in 2050**

Percentage change compared to present climate scenario

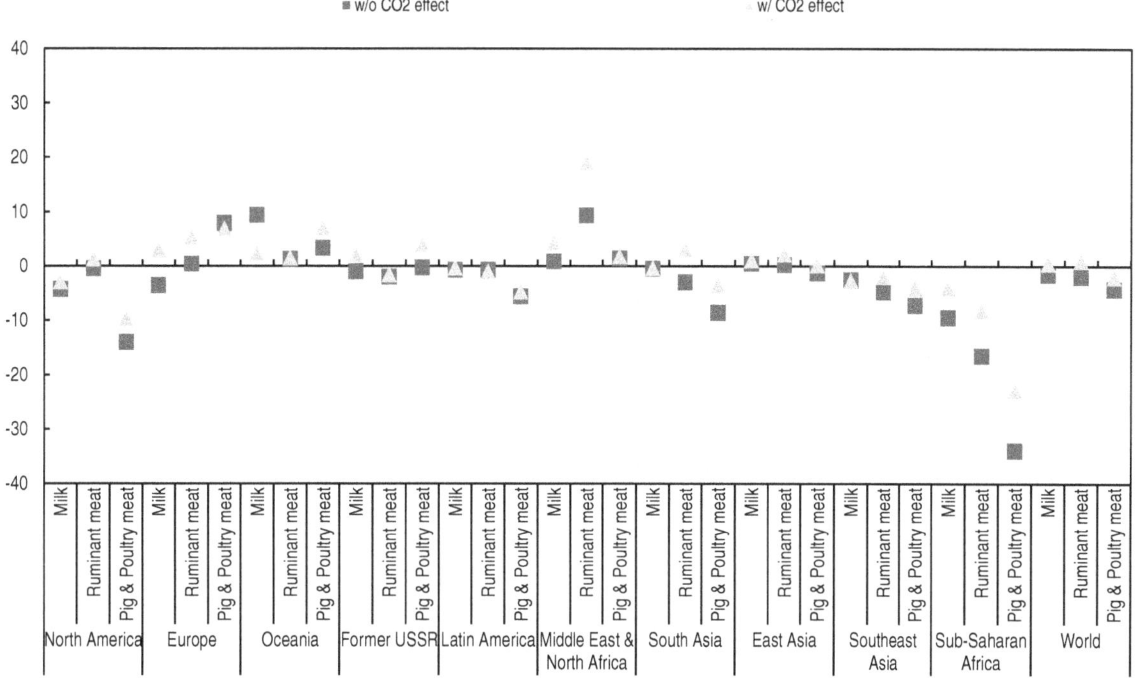

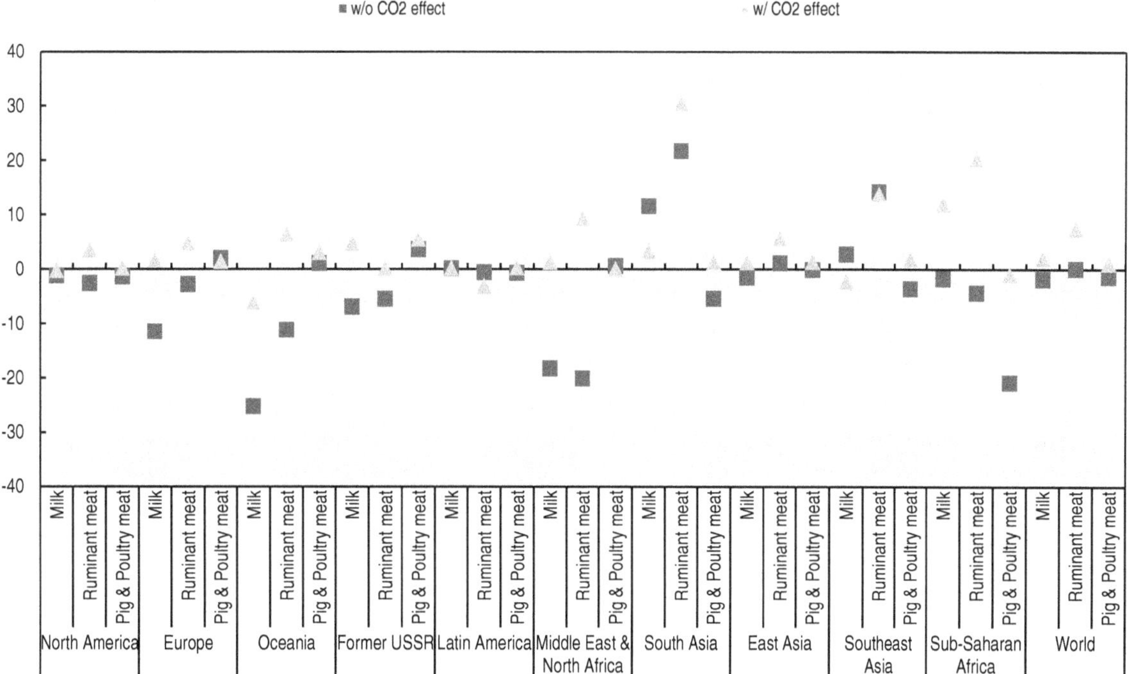

Note: EPIC and LPJmL are the underlying crop models used for the analysis.
Source: Havlík et al. (2015), Figure 8.

Pasture- and rangeland, which encompasses various different types of land that is used to keep animals (e.g. grasslands, shrublands, savannahs, hot and cold deserts, and tundra), is expected to have positive effects in some regions and negative in others. Grasslands are expected to be affected in similar ways to crop yields. In addition, the CO_2 fertilisation effect might stimulate plant growth, help plants recover from water stress events more quickly, and contribute to reduced plant mortality (IPCC, 2014a).

Very few studies have brought together and quantified the various channels through which climate change impacts pastureland and livestock simultaneously. Havlik et al. (2015) provide the most innovative analysis to date, including both impacts on grassland productivity and feedstock crop productivity; it excludes direct effects related to e.g. heat stress and diseases.[4] Their analysis uses the GLOBIOM model and follows the Representative Concentration Pathway (RCP) 8.5 climate scenario. The authors' projections show that the effects on livestock production are likely to be fairly similar to those on the crop sectors. However, as grasslands are projected to be more responsive to the benefits of climate change, and more resilient to negative impacts, than crop yields, climate change is likely to induce a shift towards more ruminants in the grazing systems. Furthermore, they find that the largest uncertainties are those on the crop model that is used and the assumed strength of the CO_2 fertilisation effect; hence different scenarios are presented for these. Finally, it highlights that grassland impacts follow very similar patterns as cropland impacts. Figure 3.4, based on Havlik et al. (2015), summarises some of main quantitative effects for the scenarios without and with a CO_2 fertilisation effect.

Similar to evidence on future climate change impacts on livestock, information on the effects of warming and other climatic drivers on *aquaculture* is limited. Pickering et al. (2011) conclude that climate change will likely be beneficial for freshwater aquaculture, except in coastal zones. No comprehensive economic study on the impacts of climate change on changes in aquaculture productivity currently exists.

3.2.2. Coastal zones

Along with impacts on physical capital and land that have been quantified and discussed in the report, sea level rise can have other market-based impacts (e.g. salt-water intrusion affecting agriculture and water supply) and affect non-market goods and services provided by coastal zones. This could be triggered by gradual increases in average sea levels, but also by changes in peak levels, including storm surges. When groundwater overdraft is causing land subsidence, sea level rise will also cause larger damages to urban areas. These non-market values are nonetheless highly prized, not surprisingly given that more than 2.5 billion people globally live in coastal areas. In addition to pure existence values, a wide variety of recreational activities and other ecosystem services are provided by coastal environments, especially by key habitats such as mangroves and coral reefs. An example where many of these issues come together is the Italian lagoon city Venice that is home to distinct culture and historical architecture.

Sea level rise is expected to have significant non-market impacts on natural habitats and landforms in coastal zones, as for example on beaches and lagoons. For Southern Californian Beaches, Pendleton et al. (2009) estimate the loss in consumer surplus caused by permanent inundation from sea level rise of 1 meter to amount to up to USD 63 million per year on average and USD 37 million in extremely stormy years. It is, however, very hard to quantify how high the related costs due to climate change are as loss of land may result in a re-allocation of various land uses, also inland.

Sea level rise might also lead to the loss of entire nation states and their distinctive cultures due to sea level rise. Low-lying island states such as the Maldives, Kiribati, Palau, and Tuvalu to be particularly at risk of being flooded. For instance, the highest point on Tuvalu is a mere 5 meters above sea level, with most land lower than one meter above sea level. The already widespread annual flooding in Bangladesh is also expected to increase due to climate change. Apart from some case studies, evidence on the magnitude of these impacts on welfare is very limited. There is still insufficient knowledge about societies' ability to adapt to rising sea levels. As is the case with non-market impacts generally, quantifying the damage to culture and other non-priced aspects of human life is extremely complex. Complete disappearance of entire islands is not a marginal effect on the local economy, and thus requires non-traditional tools to assess the economic consequences. Nonetheless, the local economic and social consequences of entire societies being deprived from their land would be enormous.

3.2.3. *Extreme events: Urban flooding*

Some of the main consequences of extreme events can be accounted for in a modelling framework, as they will affect physical capital, natural resources (e.g. land) and labour. There are however, several consequences that are related to economic activities that are difficult to integrate in the modelling framework (especially local disruptions to e.g. electricity and transport) e.g. and others that are not related to economic activities, such as anxiety, discomfort, pain and suffering, or increased mortality. Yet, quantifying such disutility effects is a challenging task. If reliable data were available for the disutility effects of extreme weather events, then valuation methods could be used to assess their importance from an economic perspective (cf. the next subsection on Health).

Global robust data for the impacts of climate change on river *floods* are not readily available. Flood risk models exist, and calculate indicators such as area at risk of flooding and population at risk of flooding. It is not straightforward to translate these indicators into factors that are directly linked to economic activity. In principle, floods have two key economic effects. First, in the vicinity of the flooded area there is a large disruption of economic activity. For one to several weeks, factories shut down, people are forced to work from other locations (thereby reducing their labour productivity), etc. Secondly, floods create permanent damages to infrastructure, buildings and such, that either need to be repaired or replaced. The former can best be approximated by looking at the affected area and affected population, while the latter is more closely linked to exposed urban assets and urban damages (which take the same information on hazards and exposure as exposed urban assets, but also includes vulnerability). Of course, floods may also cause other social impacts, for example on mental health (Stanke et al., 2012).

The GLOFRIS model (Ward et al., 2013, 2014), or more precisely the cascade of models of which GLOFRIS is a part, can be used to compare projections of future flood risks with and without climate change (Winsemius and Ward, 2015) and thus establish the additional damages due to climate change. Using the framework of Ward et al. (2013), excess urban flood damages from climate change are calculated. The model cascade first links daily projected precipitation and temperature at the half-degree grid level to daily flood volumes, and then annual maximal flood volumes. An inundation model is then linked to an impact model to establish how local inundation depth translates into expected flood impacts. By running the model cascade twice, once with current climate and once with changes in climate change (based on RCP8.5 projection and the HAdGEM climate model), respectively, the projected incremental or excess costs of climate change on flood impacts are calculated.

Figure 3.5 shows the projected urban climate damages from floods for 2080, for data that is aggregated to the country level, in order to facilitate comparison with the economic assessment of the other climate impacts. The two countries that have by far the largest projected urban flood damages are India and China. The main driver for this is the huge increase in the urban assets that are exposed in these countries. The scale of flood risks is so large in these countries that the additional damages from climate change are also huge. Bangladesh is also high in the ranking of most affected countries, but in this case the role of climate change is substantially larger. The opposite is true for e.g. Indonesia, Russia, Thailand and the main Nile countries, where flood risks are currently relatively high, but the additional damages from climate change are projected to be negative. For OECD countries, the climate-induced urban flood damages are limited to less than 50 billion USD by 2080. That does not mean that total urban flood damages, i.e. either climate-induced or not, are much smaller than in non-OECD regions. For example, the total damages by 2080 in the United States amount to 170 billion USD, in Mexico to 58 billion USD, in Germany 20 billion USD and in The Netherlands to 17 billion USD, respectively. But the climate-induced component of these damages is substantially smaller than for many non-OECD countries.

Figure 3.5. **Urban climate change damages from floods by 2080**

Billions of USD, 2005 PPP exchange rates

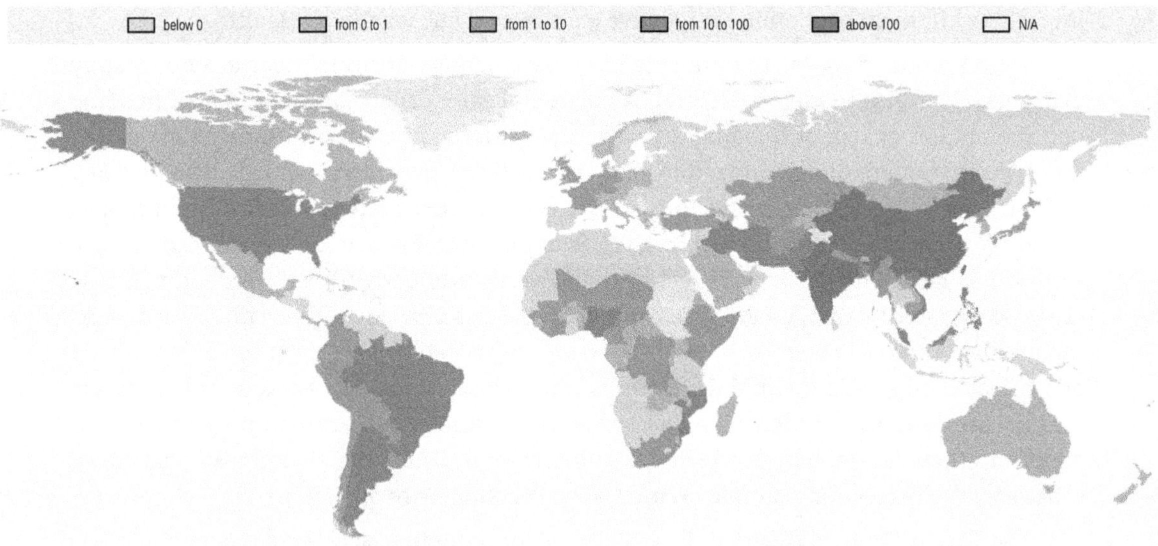

Note: This map is for illustrative purposes and is without prejudice to the status of or sovereignty over any territory covered by this map. Source of administrative boundaries: National Statistical Offices and FAO Global Administrative Unit Layers (GAUL).
Source: Own calculations based on Winsemius and Ward (2015).

Given the importance of the projected regional precipitation patterns for these simulations, and the large uncertainties surrounding them, these results are only representative for the HadGEM climate model. As Table 3.1 shows, there are significant differences when other climate models are adopted to make these projections. For instance, only the HadGEM model projections imply a reduction in urban flood damages in

Table 3.1. Climate-related potential urban flood damages by region

Billions of USD, 2005 PPP exchange rates

	HadGEM			GFDL	IPSL	MIROC	NorESM
	2010	2030	2080	2080			
OECD America							
Canada	0.0	0.5	0.0	1.5	-3.6	1.9	0.8
Chile	0.0	0.3	2.0	-3.4	-3.1	-3.4	-0.4
Mexico	0.0	-0.5	0.7	66.3	-49.7	-15.6	-29.9
USA	0.0	2.3	19.4	10.2	16.6	5.4	3.5
OECD Europe							
EU large 4	0.0	1.9	11.2	0.8	3.9	4.8	2.2
Other OECD EU	0.0	1.6	8.8	1.6	5.8	4.6	2.6
Other OECD	0.0	-0.2	1.5	-6.2	-5.2	-4.5	0.1
OECD Pacific							
Aus. and New Z.	0.0	-0.3	1.3	-4.2	1.4	1.2	0.4
Japan	0.0	0.6	3.4	2.6	1.2	0.9	1.5
Korea	0.0	0.2	0.9	1.1	0.4	2.0	0.7
OECD	0.0	6.2	49.2	70.3	-32.1	-2.8	-18.5
Rest of Europe and Asia							
China	0.0	48.0	427.9	343.0	88.8	102.5	184.4
Non-OECD EU	0.0	-0.8	-3.6	-1.7	-2.3	4.4	0.7
Russia	0.0	-5.4	-32.6	-7.6	-7.8	-4.7	-44.8
Caspian region	0.0	1.9	17.6	2.6	-4.6	2.9	-6.4
Other Europe	0.0	-2.6	-13.5	-7.9	-6.8	2.1	-12.6
Latin America							
Brazil	0.0	0.9	12.6	6.7	98.1	-15.1	-40.3
Other Lat. Am.	0.0	-0.7	15.2	-10.5	10.5	-16.6	-26.9
Middle East and North Africa							
Middle East	0.0	-0.3	39.8	-60.9	-32.2	-34.2	9.4
North Africa	0.0	-2.5	-44.9	128.0	243.2	47.2	25.0
South and South-East Asia							
ASEAN 9	0.0	-0.7	65.1	185.2	139.1	57.9	196.1
Indonesia	0.0	-2.7	-29.0	5.2	152.8	11.2	38.4
India	0.0	51.5	1 094.9	432.7	718.3	362.2	207.8
Other Asia	0.0	2.4	184.0	153.9	148.9	117.8	114.1
Sub-Saharan Africa							
South Africa	0.0	0.1	3.3	4.8	2.2	-2.0	-1.4
Other Africa	0.0	3.1	59.3	85.6	178.4	225.1	76.1
World	0.0	98.4	1 845.3	1 329.4	1 694.3	857.9	701.2

Note: HadGEM, GFDC, IPSL, MIROC and NorESM are specific climate models that are used to project precipitation and temperature patterns (see Winsemius and Ward, 2015 for more details).

Source: Own calculations based on Winsemius and Ward (2015).

StatLink http://dx.doi.org/10.1787/888933276257

Indonesia; the other models all have increased damages for this country. For the OECD region, the largest uncertainty is in the projections for Mexico. Nonetheless, there are also some consistent patterns across the models, including the fact that the largest climate-induced urban flood damages are in Asia in general and in India in particular, and that the flood damages in Russia decrease due to climate change. Using climate scenarios from the HadGEM model global urban flood damages are projected to amount to 0.7 to 1.8 trillion USD by 2080.

A complication in using these data is that they reflect *potential* damages, without any adaptive behaviour to deal with increased flood risks. Hence, the numbers presented here should be interpreted with care and are an overestimation of the least-cost urban climate damages from river floods. That said, urban damages are only one element of flood damages, and the local disruption effects are excluded here. These are likely to have severe consequences for local communities, even if their macroeconomic effect may be relatively small (cf. the damages from hurricanes).

3.2.4. Health

The modelling of health impacts in ENV-Linkages accounts for labour productivity changes due to occupational heat stress. However, it does not take into consideration premature deaths related to heat-related mortality (including heat waves), nor cold-related health effects. Non-health consequences of heat stress, such as disruptions of transport, are also not considered here.

Research on the impacts of climate change on cold-related morbidity remains scarce and inconclusive. The IPCC (2014a) states that there could be modest reductions in cold-related morbidity in some areas due to fewer cold extremes, yet it has only low confidence in this finding.

The evidence on the magnitude of the benefits of changes in premature deaths from reductions extreme cold is also mixed. According to Bosello et al. (2006) and Watkiss and Hunt (2012), the number of avoided premature deaths and the related welfare benefits of reduced winter mortality from climate change could outweigh the negative impacts from heat on mortality in certain regions. Bosello et al. (2005) project that in the European Union, the United States, Eastern European and Former Soviet Union countries, Japan, other Annex 1 countries (as defined in the United Nations Framework Convention on Climate Change), China and in India reductions in cold-related deaths from cardiovascular disease will more than offset additional deaths from heat-related and other diseases spurred by climate change in 2050. Globally, they project that climate change may lead to 849 252 fewer deaths by the middle of the century as compared to the baseline scenario. Likewise, Watkiss and Hunt (2012) find that the decrease in winter in the European Union mortality due to climate change is larger than increases in summer mortality in most of their near- to medium-term (2011-40) and long-term (2071-2100) projections. Watkiss and Hunt (2012) point to uncertainties related to the omission of extreme and urban heat island effects, however, thereby suggesting that a direct comparison of heat- with cold-related mortality in their study might not be entirely adequate.

Other studies, including by Kinney et al. (2012) and Ebi and Mills (2013), contest whether beneficial changes in winter mortality will outweigh the negative effects from increased heat-related mortality. The IPCC (2014a) also cites papers by Wilkinson et al. (2007) and regional studies by Doyon et al. (2008) and Huang et al. (2012) to conclude that "the increase in heat-related mortality by mid-century will outweigh gains due to fewer cold periods" in temperate zones and especially in tropical zones, where large populations in developing countries have limited capacity to adapt (IPCC, 2014a). Building on past empirical evidence from the United Kingdom, Staddon et al. (2014) stress that in temperate zones the link that many papers make, namely that low temperatures during winters are correlated with excess winter deaths, is empirically weak. They suggest that influenza-like illnesses – whose positive correlations with climate change remain to be proven – are the main driver for cold-related deaths. In the same vein, Honda and Ono (2009), using data from Japan, argue that risks from cold may not be ameliorated with higher average temperatures.

While the economic costs of heat-related mortality could not be accounted for in the model, they were calculated separately; this excludes any assessment of the consequences of cold-related deaths. The Japanese National Institute for Environmental Studies (NIES) and the University of Tsukuba (Japan) carried out calculations on the number of premature deaths from heat-related mortality, including heat waves. To properly align with the other projections, the RCP8.5 climate scenario is used in combination with the Hadley Centre's HadGEM climate model. Using projections of future temperature, NIES has calculated a heat index as well as an indicator of relative risk depending on temperatures. The number of additional premature deaths due to heat stress has then been calculated using the risk coefficient, baseline mortality levels as well as daily grid-level temperature data (Takahashi et al., 2007; Honda et al., 2014).[5]

The results of this analysis are presented in Table 3.2 for the ENV-Linkages regions. The regions with the highest number of premature fatalities are ones like China and India where the population is larger. Many premature deaths also take place in regions such as the EU and the US, where aging population increases the size of the vulnerable population at risk. The global death toll from heat stress is projected to increase from less than 150 thousand people annually in the current climate, to more than a million by the 2050s and close to 3 million by 2080s. However, these results do not factor in the potential for natural acclimatisation, which could reduce the number of fatalities. As regional temperatures keep rising, the number of heat stress days increases, and spells of continuous hot days get prolonged. This in turn leads to more than proportionate increases in premature deaths, in

Table 3.2. **Heat stress mortality by region**

Thousands of people

		Current climate	2030	2050	2080
OECD America	Canada	1	3	8	19
	Chile	0	1	1	3
	Mexico	1	7	12	25
	USA	11	27	63	137
OECD Europe	EU large 4	11	31	66	131
	Other OECD EU	8	22	44	75
	Other OECD	1	5	13	25
OECD Pacific	Australia and New Zealand	1	2	3	7
	Japan	3	7	10	16
	Korea	1	3	6	13
Rest of Europe and Asia	China	27	88	161	282
	Non-OECD EU	2	5	8	13
	Russia	12	20	28	40
	Caspian region	2	8	21	42
	Other Europe	5	11	16	21
Latin America	Brazil	2	8	23	58
	Other Latin America	2	9	24	100
Middle East and North Africa	Middle East	2	10	38	109
	North Africa	2	8	22	47
South and South-East Asia	ASEAN 9	2	16	39	103
	Indonesia	1	6	23	82
	India	25	55	139	369
	Other Asia	10	24	78	245
Sub-Saharan Africa	South Africa	2	3	4	12
	Other Africa	11	47	177	907
World		145	426	1 023	2 875

Source: Own calculations provided by NIES and the University of Tsukuba.

StatLink http://dx.doi.org/10.1787/888933276264

the absence of further policies. In particular, the number of premature deaths would be lower in presence of adaptive investment, including better air conditioning or wider use of early warning systems and information campaigns for the population at high risk.

The number of premature deaths per se is one indicator of the impacts to society caused by climate change. However, economists have also developed techniques to calculate the economic costs related to changes in mortality risk. Such costs allow policy makers to better evaluate the benefits of policies that would reduce the number of premature deaths. Two metrics are widely established: the value of statistical life (VSL) and the value of a life year (VOLY). Both indicators are based on the concept of society's willingness to pay for the benefits to reduce the risk of mortality. While the concept of VSL is based on the prevention of one statistical death as a whole, VOLY provides a means to account for differing lengths of remaining life expectancy (Steininger et al., 2015). Thus, the VSL approach uses a fixed value regardless of age, while the use of VOLY implies a value on people that declines with age. The VSL has been widely used in previous OECD work as it provides an evaluation of a statistical death that can be used to calculate the economic costs of environmental issues such as air pollution (OECD, 2012b and 2014b).

Controversy evolves around the fact that economists and decision-makers attach different values to changes in the risks of mortality depending on the country or region people live in. This is somewhat reflected in various studies that have attempted to estimate the VSL in different countries, as a meta-study by the OECD (2012b) shows. Figures for VSL range from USD 2 660 to USD 20 000 000 (2005 dollars) with higher values in higher income countries. Estimates also vary within countries; for instance VSL estimates for the United States vary between USD 200 000 and USD 9 400 000 (OECD, 2012b).

Despite these shortcomings, the VSL metric can be helpful to indicate the potential costs of premature deaths. It has been widely used in the context of air pollution (see e.g. OECD, 2014b), but can be applied to other cases of premature deaths, though there may be issues of context and transferability, even though the calculations are done on generic studies based on the willingness to pay of respondents to reduce the probability of premature death. Box 3.3 explains the VSL methodology more in detail.

Box 3.3. **The value of a statistical life**

Mortality risks – at the level of society as a whole – can be evaluated using the "value of statistical life" (VSL). The VSL is derived from aggregating individuals' willingness to pay (WTP) to secure a marginal reduction in the risk of premature death. OECD (2012b) analyses the empirical WTP literature on VSL and describes how to derive a VSL value from the survey. The survey finds an average value of USD 30 for a reduction in the annual risk of dying from 3 in 100 000 to 2 in 100 000, namely each individual is willing to pay USD 30 to reduce the risk of premature death by 1 in 100 000. As underlined in OECD (2012b), the VSL is not the value of an identified person's life, but an aggregation of how individuals value small changes in risk of death. As such, the economic cost of the impact being studied – which in this case is related to climate change – becomes the value of the VSL multiplied by the number of premature deaths caused.

Following a rigorous meta-analysis of VSL studies (OECD, 2012b), a set of OECD-recommended values for average adult VSL are available for OECD countries and for non-OECD G20 countries. The recommended range for OECD countries is 2005 USD 1.5 million-4.5 million, the recommended base value is USD 3 million.

The OECD (2012b) VSL methodology has been used to calculate the economic costs relative to the number of premature deaths caused by heat stress calculated by NIES and the University of Tsukuba for the OECD ENV-Linkages regions (see Table 3.3). The overall costs for OECD are of around (2010) USD 75 billion in the current climate and are projected to increase to more than USD 230 billion in 2030, almost USD 490 billion in 2050 and to over a trillion USD in 2080. The highest economic costs in 2080 are projected for North American and EU countries and particularly in the United States and the four large EU countries. Note that the use of VOLYs would imply substantially lower values.

Table 3.3. **Economic costs of premature deaths from heat stress in OECD countries using VSL**

2010 USD billions

OECD country	Current climate	2030	2050	2080
Canada	3.0	9.9	23.4	55.6
Chile	0.3	1.5	3.5	7.5
Mexico	4.0	20.3	35.6	74.3
USA	1.8	7.5	27.4	132.5
EU large 4	33.8	92.3	197.2	392.3
Other OECD EU	16.6	49.6	104.4	182.7
Other OECD	3.8	16.0	39.4	75.2
Australia and New Zealand	1.8	5.7	9.0	20.3
Japan	7.7	21.8	30.3	49.4
Korea	1.8	7.5	17.0	38.7
OECD total	***74.7***	***232.2***	***487.1***	***1 028.5***

Source: Own calculations based on number of premature deaths provided by NIES and the University of Tsukuba and on VSL values from OECD (2012b).

StatLink http://dx.doi.org/10.1787/888933276272

3.2.5. Ecosystems

Ecosystems provide a multitude of services to society and individuals, both through direct (market) and indirect (non-market) channels (TEEB, 2014). While it is challenging (and perceived as controversial by some) to account for all potential benefits of ecosystem services, not considering ecosystem services is likely leading to a significant understatement of the costs of ecosystem damage from climate change. Among these services, there are many that humans do not actively use for economic purposes, but which they appreciate for other reasons.

The risk of producing biased estimations from neglecting non-use values might be especially present when assessing the value of *biodiversity* to society, which can be defined as the "variability among living organisms from all sources including, inter alia, terrestrial, marine and other aquatic ecosystems and the ecological complexes of which they are a part" (UN, 1992). Few comprehensive studies exist that have valued the costs of biodiversity loss from climate change to society on a global scale. The OECD's *Environmental Outlook to 2050* projects that climate change will contribute roughly 40% of the additional loss of terrestrial mean species abundance between 2010 and 2050 in the report's baseline scenario, representing the largest of all drivers (Figure 3.6; reproduced from OECD, 2012a). This makes climate change the largest driver of biodiversity loss, with stronger impacts than those related to food crops, bioenergy, pasture, forestry, former land use, nitrogen, and infrastructure, encroachment and fragmentation (including urbanisation).

Figure 3.6. **Climate change adds pressure to biodiversity loss**

Relative shares of various pressures to additional terrestrial loss in mean species abundance

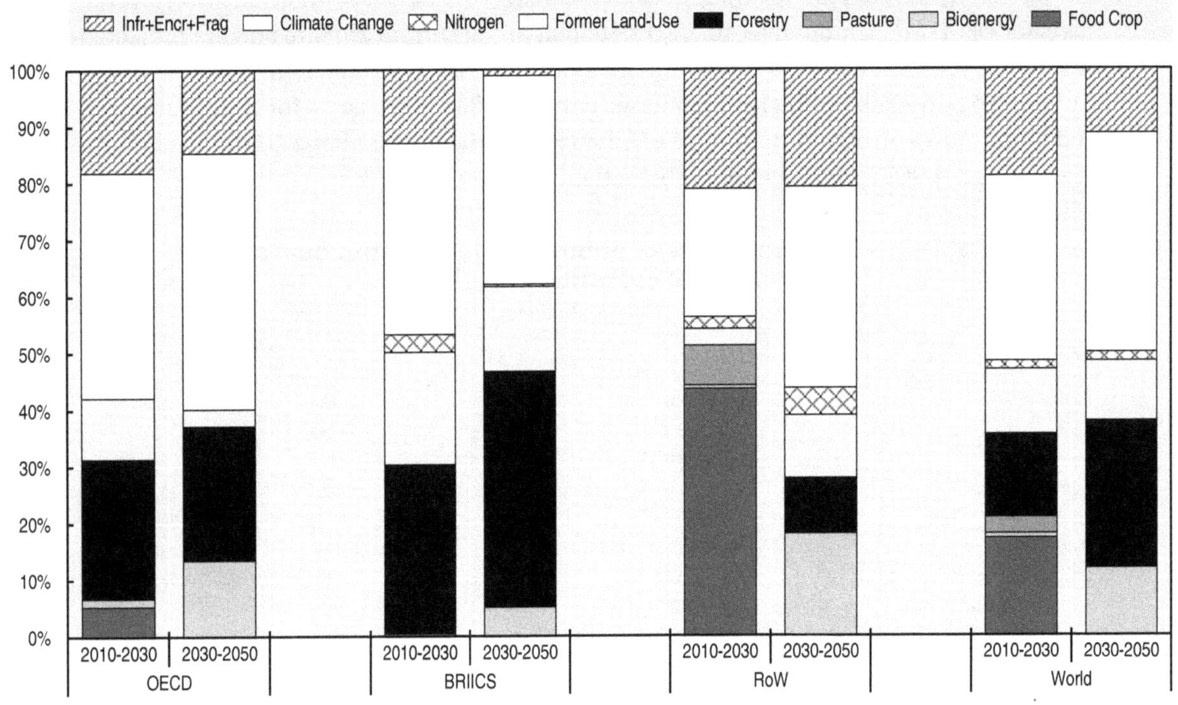

Source: IMAGE model, as reported in OECD (2012a).

StatLink http://dx.doi.org/10.1787/888933276101

More generally, the projection of valuation of *losses in biodiversity and ecosystem services* tries to reflect the change in value of biodiversity and ecosystem services due to climate change. Ecosystems do not usually have a market value and it is difficult to attribute one to them. In terms of the CGE modelling framework the reliance of economic sectors on these services is also poorly understood. Following Bosello et al. (2012), an economic valuation of ecosystem services is therefore based on a modified willingness to pay (WTP) approach. The initial assumption is that these services are largely non-marketed and not directly marketable. Accordingly, their value can only be extracted through elicitation of preferences. In particular the WTP to avoid a given loss in ecosystems is used to approximate the lost value in case they are not protected. This is for instance the methodology applied in Stanford University's MERGE model (Manne et al., 1995). Using this setup, and as explained in more detail in Warren et al. (2006), an equation to link WTP to temperature increases is established, using Eurostat data on expenditures of 0.62% of GDP in the EU, which is (boldly) assumed to protect against 2°C of warming. A logistic function is then used to calculate WTP for different countries in different periods based on the regional GDP pathways (for the no-damage baseline projection). The (bold) implicit assumptions are that what is actually paid is reasonably close to the WTP, and roughly sufficient to preserve ecosystems and their services in a world with moderately increasing temperatures. Of course, this provides only indirect information on the value of the damages of the loss of ecosystem services to the economy; how specific sectors will be (or already are) affected by loss of biodiversity and ecosystem services cannot be inferred from this assessment.

As shown in Table 3.4, by 2060, the regions with the highest WTP, which is of around 1% of GDP in the RCP8.5 scenario, are large economies such as Japan, Korea, the United States, Canada, Mexico, South Africa as well as many European countries. The WTP is smaller (between 0.3% and 0.7% of GDP) in Chile and the rest of Latin America, China, Russia, Middle East, and in OECD EU regions. Other regions have very small WTP, with the smallest being in the group of Sub-Saharan African countries. This distribution of WTP values is not surprising as the willingness to pay for ecosystem services is assumed to be directly linked to average income. Hence, it is natural that the WTP for ecosystem services is higher in richer countries such as the United States or Canada while hardly existing in other areas of the world such as continental Africa. For RCP6.0, temperature increases until 2060 are smaller, and hence the pressure on ecosystems is less pronounced, leading to smaller values.

Table 3.4. **Willingness to pay for ecosystem service conservation by region**

Percentage of GDP in 2060

		RCP6.0	RCP8.5
OECD America	Canada	0.5	1.1
	Chile	0.3	0.6
	Mexico	0.4	0.9
	USA	0.5	1.1
OECD Europe	EU large 4	0.5	1.1
	Other OECD EU	0.5	1.1
	Other OECD	0.5	1.1
OECD Pacific	Australia and New Zealand	0.5	1.1
	Japan	0.5	1.1
	Korea	0.5	1.1
Rest of Europe and Asia	China	0.2	0.5
	Non-OECD EU	0.3	0.7
	Russian Federation	0.2	0.4
	Caspian region	0.1	0.1
	Other Europe	0.1	0.2
Latin America	Brazil	0.1	0.2
	Other Latin America	0.1	0.3
Middle East and North Africa	Middle East	0.1	0.3
	North Africa	0.1	0.1
South and South-East Africa	ASEAN 9	0.1	0.1
	Indonesia	0.0	0.1
	India	0.0	0.1
	Other Asia	0.0	0.1
Sub-Saharan Africa	South Africa	0.4	0.8
	Other Africa	0.0	0.0

Source: Own calculations based on Warren et al. (2006).

StatLink http://dx.doi.org/10.1787/888933276283

Besides non-market impacts, some other market activities that rely on functioning ecosystems will likely be significantly impacted by climate change. *Forestry* is such an example. The OECD's *Environmental Outlook to 2050* projects that the global area of production forests (forests managed for the production of timber, pulp and paper, and fuelwood) will increase by almost 60% between 2010 the mid of the century under the report's Baseline scenario, reaching a total size of 15 million km^2 (OECD, 2012a). In contrast, primary (unmanaged) forests are projected to continue to decline. Changes in the productivity of production forests depend on a variety of factors, such as location, tree species, water availability, and the effects of CO_2 fertilization. Successful adaptation will lead to increased forest plantation productivity among the majority of producers. Producer benefits will vary with latitude and region, however: only low to mid latitude producers are expected to benefit from climate change, while mid to high latitude producers will be hurt by lower market prices (IPCC, 2014a). On a global level benefits from physical changes from climate change to forestry are estimated to outweigh the costs (Kirilenko and Sejo, 2007; IPCC, 2014a). However, none of these studies has carried out economic evaluations to assess the size of the global market impact of climate change on forestry.

3.2.6. *Water stress*

Stress on freshwater availability induced by higher temperatures, changed precipitation patterns, glacier melting and other climatic drivers is expected to have economic consequences for water-intensive economic activities, such as for irrigation in agriculture or for cooling in energy supply. Negative effects are especially likely for many dry subtropical regions. In other regions, the net availability of water may not be affected or increase. While water-intensive sectors are the most directly affected, reduced water availability from climate change will affect households and municipalities through impacts on the availability and quality of drinking water (OECD, 2013, 2014a; IPCC, 2014a), although water allocation rules crucially affect how water stress affects different parts of the economy (OECD, 2015).

Extensive work on the impact of climate change on global water resources exists (e.g. Schewe et al., 2013; Hejazi et al., 2014; Fung, et al., 2011; Alcamo, 2007; Arnell, 2004), yet the literature on the economic consequences of these impacts is not yet as well developed. A few studies have explored the likely level of damage both globally and for certain regions; some have quantified the adverse effects from climate change to specific water usages (e.g. irrigation), while others have chosen a more comprehensive approach, assessing the costs on a multitude of water usages. The majority of these studies suggest substantial costs from climate change (IPCC, 2014a). Most studies focus on the impacts of water stress on crop yields, which is already accounted for in the agricultural damages category. Studies on specific water impacts not related to agricultural uses are much more scarce.

Empirical contributions also highlight the increased cost of reduced water availability from climate change for supplying electricity. Under higher temperatures, the efficiency, output and reliability of thermal power plants, for example, is expected to suffer as a consequence of reduced water volume and higher water temperature – two factors that are crucial for appropriate cooling of most existing plants (alternative processes, such as dry cooling, typically consume more electricity and require higher investment costs; similarly, alternative placement of the power plant to more water-secure places, such as next to the sea, tends to increase the costs of transporting power to the users). Climate change might considerably raise the costs of power plant operation if climatic drivers accelerate water scarcity in these areas. China might be particularly affected by this development given that much of the existing and planned coal power capacity is located in regions with high risks of water stress. Cost increases in India, in turn, are expected to be much smaller given that Indian coal mines, power stations and industrial demand are mostly located in areas with low risks of water scarcity (IEA, 2015; WRI, 2014). A case study by Hurd et al. (2004) has assessed the likely welfare costs of climate change impacts on water use in electric power generation in the United States, projecting losses of about USD 622 million per year up to 2100 due to changes in cooling water for combustion in coal, natural gas and other thermal power stations. The study assumed warming of +2.5°C above pre-industrial levels and a drop of 10% in monthly average precipitation. Water shortage can also negatively affect the operation of hydropower plants (IPCC, 2014a).

A recent study by Henderson et al. (2013) attempted to estimate the economic impacts of climate change on water resources in the United States, covering several other types of water use beyond irrigation and cooling. They suggest annual damages of approximately USD 2.1 billion by 2050 and USD 4.2 billion by 2100 without new climate change policies. The largest impacts are projected to affect non-consumptive activities, such as hydropower and environmental flows; agriculture and other consumptive uses will be impacted by climate change less negatively. Similarly, Strzepek et al. (2014) suggest negative welfare consequences for the United Sates in the order of USD 6.5 to 15 billion by the end of the century in their assessment of the impacts of climate change on water supply, management, and use of water resources. For the year 2050, results are more ambiguous, with one scenario suggesting positive effects on welfare from climate change and two others negative effects.

3.2.7. Human security

Human migration

The flows, magnitude and forms of human migration, which describes "a permanent or semi-permanent move by a person for at least one year that involves crossing an administrative, but not necessarily a national border" (Brown and Bean, 2005, as referenced in IPCC, 2014a), are likely going to be affected by climate change (IPCC, 2014a). The review by the International Organization for Migration suggests that climate change may displace between 200 and 250 million people within countries or to other jurisdictions by 2050 (IOM, 2009; Shamsuddoha and Chowdhury, 2009). None of these or other estimates are very reliable, however, and e.g. Tacoli (2009), Bettini (2013) and McAdam (2011) argue that robustly linking climate change to migration is problematic. Also, the effect of climate change on migration can be seen through two lenses: as a forced displacement of people whose livelihood is threatened, or as a powerful adaptation strategy to limit the potential welfare consequences of climate change impacts (Waldinger, 2015). In its Fifth Assessment Report, the IPCC does not cite any global estimates on climate change-induced migration, apart from the impact of sea level rise on displacement. World Bank (2011, 2012a, 2012b) include a comprehensive literature review of how to appropriately account for the costs of forced displacement, and provides general guidelines for assessment.

Despite uncertainty about the size of the impact, researchers have nevertheless aimed to explore the possible linkages between climate change and migration, among others by studying past weather events and their impact on human movement. On this basis, the literature indicates that the projected increase in frequency and intensity of certain types of extreme weather events due to climate change will likely be the major climate change-related drivers for forced human displacement in the future (IPCC, 2014a). Extreme weather events have already forced people to leave their homes in the past, and if their frequency or intensity increases due to climate change, this will put an additional pressure on displacement. Short-term, local disruptions do, however, not necessarily lead to permanent migration (OECD, 2014c; IPCC, 2014a; Fussell et al., 2010). Gradual changes in regional climate conditions can also contribute to migration, yet this linkage is not well documented in the literature.

Empirical evidence of the economic impact of climate change-related migration, both from extreme weather events and gradual climate change, is very limited, however – above all because of the uncertainty surrounding the exact relationship between climate change and human migration. So far, most studies have focused on qualitatively investigating the various reasons and risks that might push people towards leaving their homes. One important distinction the literature draws is that migration induced by climate change might well be different from migration for labour purposes in its consequences for the economy. Migration during or after catastrophic climate events could put large pressure on host region infrastructure and other services, in a short period of time. It could also have significant negative effect on labour force availability in the sending region, especially when disruptive.

Civil conflict

While violent conflicts can have multiple causes, poverty and economic shocks are some characteristic factors that frequently play a role in the onset or intensification of conflicts. Hsiang et al. (2013) aim to identify patterns in the relationship between climate change and civil conflict among 60 quantitative empirical studies, and find that climate change exacerbates these causes. Other studies, in turn, have not been able to establish such an association, thereby leading the IPCC to conclude only limited empirical evidence that a relationship between warming and conflict exists (IPCC, 2014a; Gleditsch and Nordås, 2014; Buhaug et al., 2012). Scientists also disagree whether climate change will be a likely direct cause for warfare between states if it were to amplify rivalry about natural resources such as water (IPCC, 2014a). There is, however, high agreement and robust evidence that climate change will have negative consequences for human security that is unrelated to civil conflict (IPCC, 2014a).

3.2.8. Tipping points

Various large-scale singular events, as the IPCC (2014a) calls them, or tipping points, in the terminology of e.g. Lenton et al. (2008), pose a systemic risk as they have far-reaching consequences that go beyond individual countries and can substantially affect the global economy. Kriegler et al. (2009) use expert elicitation of climate experts to find some consensus that there is a non-negligible probability of at least one major event taking place, even at relatively low levels of carbon concentrations. The probability that this will take place before 2060 is uncertain and likely to be small, with the exception of artic sea ice loss, but the build-up of greenhouse gases in the atmosphere in the coming decades contributes to an increased risk of crossing irreversible tipping points that trigger these events.

Figure 3.7 contains a stylised overview of some of the major risks for large-scale tipping points, based on a 2014 update of Lenton et al. (2008). The probability of these different events varies widely, as does the time-scale over which these systems would collapse. They are, however, not independent of one another: in some cases these events may stimulate each other, but in other cases they slow each other down.

The non-linear effects that are caused by these discontinuities or tipping points are highly uncertain, but such disruptive changes would induce a major shock to both the climate and the economic system, albeit often with large regional differences. In the quantitative assessments of the impacts of climate change, catastrophic risks are mostly ignored (Lenton and Ciscar, 2013). Nonetheless, many authors have claimed that catastrophic risks may be much more important than the more gradual changes that have been assessed in the analysis above (e.g. Pindyck, 2013; Stern, 2013, Weitzman, 2013).

Figure 3.7. **Regional tipping points**

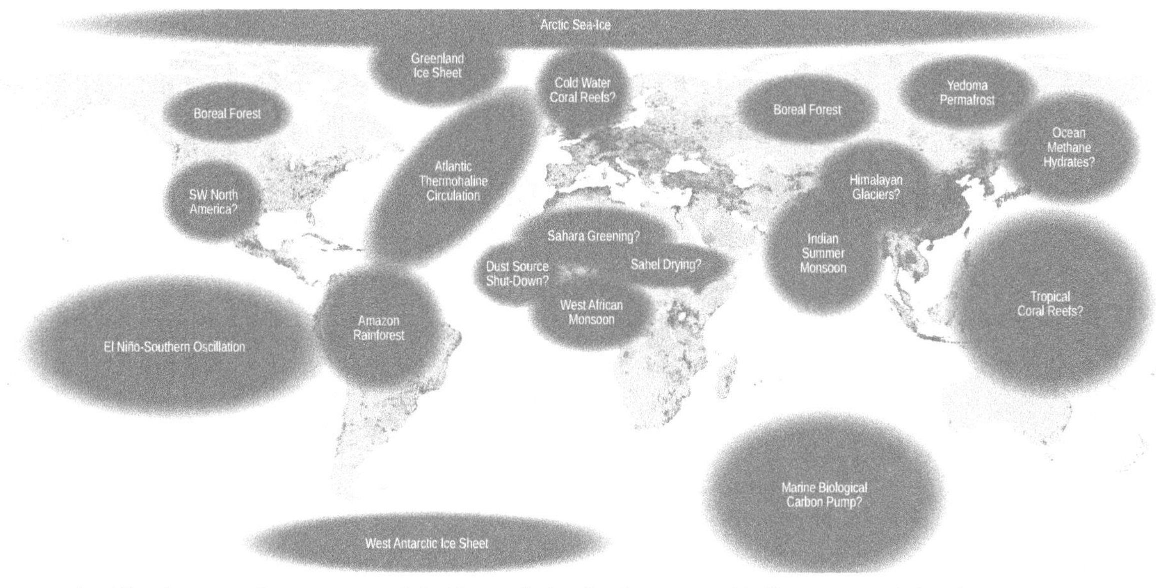

Source: Updated by Tim Lenton from Lenton et al. (2008); population density map provided by CIESIN et al. (2005).

Some applied modelling studies adopt an *ad hoc* approach and make an assumption of a permanent fixed percentage loss of regional or global GDP; e.g. Hope (2006) chooses 10% of GDP for the EU and varying rates for other countries, with China the lowest at 2% and India the highest at 25%; Nordhaus (2007) chooses 25% of GDP, but does not specify a source for this. Bosello et al. (2014) also adopt the ad-hoc assumption of 25% GDP loss when a catastrophe occurs, but assume it replaces the gradual damage function; the associated expected damage (i.e. including the probability that the catastrophe occurs) they thus calculate is between 6 and 14 per cent of GDP by 2100. The damage function with a high power term on temperature as presented in Weitzman (2012) and discussed above in the context of the AD-DICE projections, can also be seen as a way to embody such catastrophic risks; the difference in GDP loss between the damage functions of Nordhaus and Weitzman amounts to just over 10% of global GDP. Effectively, there is little evidence on which to base the economic impact of catastrophic risks. A further complication is that the economic modelling tools, including computable general equilibrium (CGE) models such as ENV-Linkages, are based on a marginal approach: large shocks change behaviour in ways that are not captured by the smooth elasticity-based functions in the modelling frameworks.

While the future economic costs of catastrophic events are difficult to assess, the risks imposed by such catastrophes can be approximated through a hazard function (Gjerde et al., 1999); that is, the chance that no major catastrophe occurs in the current period, given that none has occurred in the past. There is insufficient information to robustly calibrate such a hazard function; the recalibration presented here matches the more recent information in Kriegler et al. (2009), which is based on an expert elicitation. The same methodology is used in Bosello et al. (2014) and Lontzek et al. (2015). The hazard function drawing on the expert elicitation suggests that the chance of at least one of these catastrophic events being irreversibly triggered (though probably not fully deployed) by 2060 could be as large as 16% in the central projection, i.e. the hazard rate of not triggering any catastrophic event declines to 84% (Figure 3.8).[6] This assessment is unfortunately not updated to the latest scientific findings, such as those on arctic sea ice loss, where a seasonally ice-free Arctic

Figure 3.8. **Hazard rate of catastrophic events**

Chance of not triggering any catastrophic event

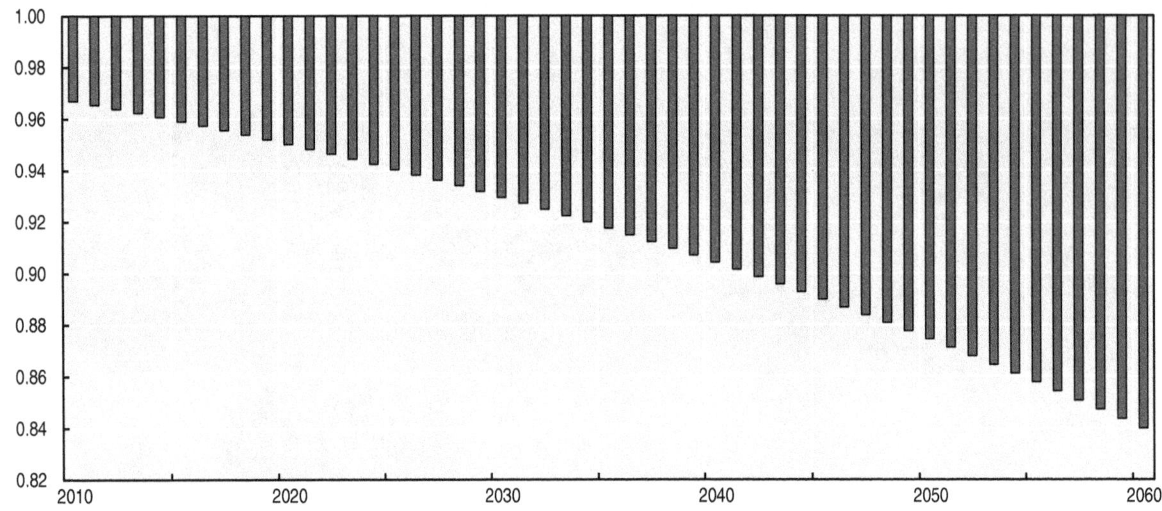

Source: Own calculations based on Kriegler et al. (2009).

StatLink http://dx.doi.org/10.1787/888933276118

ocean is likely by mid-century, and where there is little evidence for a specific tipping point (IPCC, 2013). Given the large economic consequences of such events, this probability can be interpreted as a risk premium or option value that should be placed on current emissions, reflecting their long-term potential implications. A robust quantification of such a risk premium, with regional differentiation, should be a high research priority for climate economists. Cai et al. (2015) and Markandya et al. (2015) provide an excellent starting point for this; using a stochastic modelling approach, both studies find that the risks of crossing tipping points leads to a substantial increase in the benefits of policy action. Markandya et al. (2015) also highlight the critical role of the degree of risk aversion, arguing that a high degree of risk aversion could imply a risk premium of around 100% on the social cost of carbon, while a risk-neutral approach yields a risk premium of around 10%. Thus, despite their uncertainty, ambitious mitigation action is warranted to reduce the risks of crossing the tipping points and avoid locking in irreversible climate change.

Notes

1. As AD-DICE is a forward-looking model, current GDP levels depend on future damage levels. Therefore, the GDP losses are calculated for each scenario separately as percentage of GDP. Alternatively, one could assume fixed savings rates, but that goes against the philosophy of the model that aims at identifying least-cost pathways.
2. Eliminating short-lived gases that have a cooling effect, such as sulphate aerosols, at the same time would actually lead to a temporary warming of "a few tenth of a degree" (IPCC, 2013).
3. The interested reader might find a more comprehensive review of the various implications of climate change on economic activity and society, although often less detailed on quantitative aspects, in the contribution of Working Group II of the IPCC's Fifth Assessment Report (IPCC, 2014a).
4. The inclusion of crop productivity effects represents a double-counting with respect to the agricultural impacts presented in Chapter 2 and precludes a direct incorporation of these effects in ENV-Linkages.
5. There is a discussion in the literature over the extent to which these premature deaths represent short-term displacement mortality ("harvesting"), i.e. people that die from heat stress may have serious

existing health conditions or are very old, i.e. such that that the period of life lost is small. Following Honda et al. (2014), this is (crudely) taken into account in the calculations through a lag term.

6. And it follows from the inertia in the climate system that this risk would remain intact for more than a century, even if emissions were to drop to zero immediately after 2060.

References

Ahmed, M., C.K. Chong and H. Cesar (2005). *Economic Valuation and Policy Priorities for Sustainable Management of Coral Reefs, Second Edition*, WorldFish Center: Penang, Malaysia.

Alcamo, J., M. Flörke, and M. Märker (2007), "Future long-term changes in global water resources driven by socio-economic and climatic changes", *Hydrological Sciences Journal*, Vol. 52(2), pp. 247-275.

Arnell, N.W. (2004), "Climate change and global water resources: SRES emissions and socio-economic scenarios", *Global Environmental Change*, Vol. 14(1), pp. 31-52.

Bettini, G. (2013), "Climate barbarians at the gate? A critique of apocalyptic narratives on 'climate refugees'", *Geoforum*, Vol. 45, pp. 63-72.

Bosello, F., E. de Cian and L. Ferranna (2014), "Catastrophic risk, precautionary abatement, and adaptation transfers", *FEEM Nota di Lavoro*, No. 108.2014, Milan.

Bosello, F., R. Roson and R.S.J. Tol (2006), "Economy-wide estimates of the implications of climate change: Human health", *Ecological Economics*, Vol. 58, pp. 579-591.

Brown, S. and F. Bean (2005), "International migration", in *Handbook of Population* [Poston, D.L. and M. Micklin (eds.)], Kluwer, Dordrecht, Netherlands, pp. 347- 382.

Buhaug, H. and O.M. Theisen (2012), "On environmental change and armed conflict", in *Climate Change, Human Security and Violent Conflict: Challenges for Societal Stability* [Scheffran, J., M. Brzoska, H.G. Brauch, P.M. Link, and J. Schilling (eds.)]. Springer-Verlag Berlin Heidelberg, Germany, pp. 43-55.

Buhaug, H., N.P. Gleditsch and O.M. Teishn (2008), "Implications of climate change for armed conflict", paper presented at World Bank Group workshop on Social Dimensions of Climate Change, Washington, DC, 5-6 March 2008, *http://siteresources.worldbank.org/INTRANETSOCIALDEVELOPMENT/Resources/SDCCWorkingPaper_Conflict.pdf*.

Cai, Y., K.L. Judd, T.M. Lenton. T.S. Lontzek and D. Narita (2015), "Environmental tipping points significantly affect the cost-benefit assessment of climate policies", *Proceedings of the National Academy of Sciences of the United States of America (PNAS)*, Vol. 112, pp. 4606-4611.

Center for International Earth Science Information Network – CIESIN – Columbia University, and Centro Internacional de Agricultura Tropical – CIAT (2005), *Gridded Population of the World, Version 3 (GPWv3): Population Density Grid (Population Density 2000)* [Map]. Palisades, NY: NASA Socioeconomic Map and Applications Center (SEDAC), *http://sedac.ciesin.columbia.edu/data/set/gpw-v3-population-density/maps/services* (accessed 27 August 2015).

Ciscar. J.C. et al. (2014), "Climate Impacts in Europe. The JRC PESETA II Project", *JRC Scientific and Policy Reports*, No. EUR 26586EN, Luxembourg: Publications Office of the European Union.

Costanza, R., M. Hart, S. Posner and J. Talberth (2009), "Beyond GDP: The need for new measures of progress", *The Pardee Papers*, No. 4 (January 2009), Boston University.

De Bruin, K.C. (2014), "Calibration of the AD-RICE2012/AD-DICE2013 models", *CERE Working Paper*, No. 2014:3, Center for Environmental and Resource Economics, Umeå.

Dell, M., B.F. Jones and B.A. Olken (2012), "Temperature shocks and economic growth: Evidence from the last half century", *American Economic Journal: Macroeconomics*, Vol.4(3), pp. 66-95.

Dietz, S. and N. Stern (2015), "Endogenous growth, convexity of damages and climate risk: how Nordhaus' framework supports deep cuts in carbon emissions", *The Economic Journal*, forthcoming; available as Grantham Research Institute on Climate Change and the *Environment Working Paper* 159, London.

Doyon, B., D. Belanger and P. Gosselin (2008), "The potential impact of climate change on annual and seasonal mortality for three cities in Quebec, Canada", *International Journal of Health Geographics*, Vol. 7(23).

Ebi, K.L. and D. Mills (2013), "Winter mortality in a warming climate: A reassessment", *WIREs Climate Change*, Vol. 4, pp. 203-212.

EC (2009a), *GDP and beyond: Measuring progress in a changing world*, Communication from the Commission to the Council and the European Parliament, COM(2009) 433 final.

EC (2009b), "Summary notes from the Beyond GDP conference", summary notes on the Beyond GDP conference organised by the European Commission, European Parliament, Club of Rome, OECD, and WWF, Brussels, 19-20 November 2007, *http://ec.europa.eu/environment/beyond_gdp/proceedings/bgdp_proceedings_summary_notes.pdf*.

Felbermayr, G. and J. Gröschl (2014), "Naturally negative: the growth effects of natural disasters", *Journal of Development Economics* Vol. 111, pp. 92-106.

Fung, F., A. Lopez and M. New (2011), "Water availability in +2°C and +4°C worlds", *Philosophical Transactions of the Royal Society A: Mathematical, Physical and Engineering Sciences*, Vol. 369, pp. 99-116.

Fussell, E., N. Sastry and M. Vanlandingham (2010), "Race, socioeconomic status, and return migration to New Orleans after Hurricane Katrina", *Population and Environment*, Vol. 31 (1-3), pp. 20-42.

Gjerde, J., S. Grepperud and S. Kverndokk (1999), "Optimal climate policy under the possibility of a catastrophe", *Resource and Energy Economics*, Vol. 21, pp. 289-317.

Gleditsch, N.P. and R. Nordas (2014), "Conflicting messages? The IPCC on conflict and human security", *Political Geography*, Vol. 43, pp. 82-90.

Government Office for Science (2011), *Migration and Global Environmental Change: Future Challenges and Opportunities*, Final Project Report, London.

Graux, A.-I. et al. (2011), "Development of the pasture simulation model for assessing livestock production under climate change", *Agriculture, Ecosystems and Environment*, Vol. 144, pp. 69-91.

Havlík, P.D. et al. (2015), "Global climate change, food supply and livestock production systems: A bioeconomic analysis", in *Climate Change and Food Systems: Global Assessments and Implications for Food Security and Trade* [Aziz Elbehri (ed.)], Food Agriculture Organization of the United Nations (FAO), Rome.

Hejazi, M.I. et al. (2014), "Integrated assessment of global water scarcity over the 21st century under multiple climate change mitigation policies", *Hydrology and Earth System Sciences*, Vol. 18(8), pp. 2859-83.

Henderson, J. et al. (2015), "Economic impacts of climate change on water resources in the coterminous United States", *Mitigation and Adaptation Strategies for Global Change*, Vol. 20 (1), pp. 135-157.

Honda, Y. et al. (2014), "Heat-related mortality risk model for climate change impact projection", *Environmental Health Prevention Medicine*, Vol. 19, pp. 56-63.

Honda, Y. and M. Ono (2009), "Issues in health risk assessment of current and future heat extremes", *Global Health Action*, Vol. 2. 10.3402/gha.v2i0.2043.

Hope, C. (2006), "The marginal impact of CO_2 from PAGE2002: An integrated assessment model incorporating the IPCC's five reasons for concern", *The Integrated Assessment Journal*, Vol. 6(1), pp. 19-56.

Hsiang, S., M. Burke and E. Miguel (2013), "Quantifying the influence of climate on human conflict", *Science*, Vol. 341 (6151).

Hsiang, S.M. and A.S. Jina (2014), "The causal effect of environmental catastrophe on long-run economic growth", *NBER Workshop Paper*, No. 20352.

Huang, C. et al. (2012), "The impact of temperature on years of life lost in Brisbane, Australia", *Nature Climate Change*, Vol. 2(4), pp. 265-270.

Hurd. B.H. et al. (2004), "Climatic change and U.S. water resources: From modeled watershed impacts to national estimates", *Journal of the American Water Resources Association*, Vol. 40 (1), pp. 129-148.

IEA (2015), *World Energy Outlook Special Report on Energy and Climate Change*, International Energy Agency, Paris.

IOM (2009), *Migration, Environment and Climate Change: Assessing the Evidence*, International Organization for Migration (IOM), [Laczko, F. and C. Aghazarm (eds.)], Geneva, Switzerland.

IPCC (2014), *Climate Change 2014: Impacts, Adaptation, and Vulnerability. Part A: Global and Sectoral Aspects*, Contribution of Working Group II to the Fifth Assessment Report of the Intergovernmental Panel on Climate Change [Field, C.B., V.R. Barros, D.J. Dokken, K.J. Mach, M.D. Mastrandrea, T.E. Bilir, M. Chatterjee, K.L. Ebi, Y.O. Estrada, R.C. Genova, B. Girma, E.S. Kissel, A.N. Levy, S. MacCracken, P.R. Mastrandrea, and L.L.White (eds.)]. Cambridge University Press, Cambridge, United Kingdom and New York, NY, USA, 1132 pp.

IPCC (2013), *Climate Change 2013: The Physical Science Basis*, Contribution of Working Group I to the Fifth Assessment Report of the Intergovernmental Panel on Climate Change [Stocker, T.F., D. Qin, G.-K. Plattner, M. Tignor, S.K. Allen, J. Boschung, A. Nauels, Y. Xia, V. Bex and P.M. Midgley (eds.)]. Cambridge University Press, Cambridge, United Kingdom and New York, NY, USA, 1535 pp.

IPCC (1996), *Climate Change 1995: The Science of Climate Change*, Contribution of Working Group I to the Second Assessment Report of the Intergovernmental Panel on Climate Change [Houghton, J.T., L.G. Meira Filho, B.A. Callander, N. Harris, A. Kattenberg, and K. Maskell (eds.)]. Cambridge: Cambridge University Press.

Kinney, P.L. et al. (2012), "La mortalité hivernale va-t-elle diminuer avec le changement climatique ?" [Winter mortality in a changing climate: will it go down?], *Bull Epidémiol Hebd*, Vol. 12-13, pp. 148-151.

Kirilenko, A.P. and R.A. Sedjo (2007), "Climate change impacts on forestry", *Proceedings of the National Academy of Sciences (PNAS)*, Vol. 104 (50), pp. 19697-702.

Kriegler, E. et al. (2009), "Imprecise probability assessment of tipping points in the climate system", *Proceedings of the National Academy of Science (PNAS)*, Vol. 106(13), pp. 5041-46.

Lenton, T. and J.C. Ciscar (2013), "Integrating tipping points into climate impact assessments", *Climatic Change*, Vol. 117, pp. 585-597.

Lenton, T. et al. (2008), "Tipping elements in the Earth's climate system", *Proceedings of the National Academy of Sciences (PNAS)*, Vol. 105(6): 1786-93.

Lontzek et al. (2015), "Stochastic Integrated Assessment of Climate Tipping Points Calls for Strict Climate Policy", *Nature Climate Change*, forthcoming.

Loomies, J.B. and D.S. White (1996), "Economic benefits of rare and endangered species: Summary and meta-analysis", *Ecological Economics*, No. 16, pp. 197-206.

Manne, A., R. Mendelsohn and R. Richels (1995), "MERGE – A model for evaluating regional and global effects of GHG reduction policies", *Energy Policy*, Vol. 23(1), pp. 17-34.

Markandya, A., E. De Cian, F. Bosello, J. Polanco and L. Drouet (2015), " Building Uncertainty into the Adaptation Cost Estimation in Integrated Assessment Models", *BASE Project Working Paper*, BC3, Bilbao.

Mayeres I. and D. Van Regemorter (2008), "Modelling the health related benefits of environmental policies and their feedback effects: A CGE analysis for the EU countries with GEM-E3", *Energy Journal*, Vol. 29, pp. 135-50.

McAdam, J. (2011), "Refusing refuge in the Pacific: deconstructing climate-induced displacement in international law", in *Migration and Climate Change* [Piguet, E., A. Pécoud and P.d. Guchtenreise eds.], United Nations Educational, Scientific and Cultural Organization (UNESCO), Cambridge University press, Cambridge, UK and New York, USA, pp. 102-137.

McLeman, R.A. (2014), "Climate and human migration – past experiences", *Future Challenges*, Cambridge: Cambridge University Press.

Millennium Ecosystem Assessment (2005), *Ecosystems and Human Well-being: Biodiversity Synthesis*, World Resources Institute, Washington, DC.

Moore, F.C. and D.B. Diaz (2015), "Temperature impacts on economic growth warrant stringent mitigation policy", *Nature Climate Change*, Vol. 5, pp. 127-131.

Nordhaus, W.D. (2012), "Integrated Economic and Climate Modeling", in: P. Dixon and D. Jorgenson (eds.), *Handbook of Computable General Equilibrium Modeling*, Elsevier.

Nordhaus, W.D. (2007), *A question of balance*, Yale University Press, New Haven, United States.

OECD (2015), *Water Resources Allocation: Sharing Risks and Opportunities*, OECD Studies on Water, OECD Publishing, Paris, *http://dx.doi.org/10.1787/9789264229631-en.*

OECD (2014a), "Climate Change, Water and Agriculture: Towards Resilient Systems", *OECD Studies on Water*, OECD Publishing, Paris, *http://dx.doi.org/10.1787/9789264209138-en.*

OECD (2014b), *The Cost of Air Pollution: Health Impacts of Road Transport*, OECD Publishing, Paris, *http://dx.doi.org/10.1787/9789264210448-en.*

OECD (2014c), "Perspectives on Global Development 2016: New Challenges and Opportunities for International Migration in a Shifting World", Concept Note for Expert Meeting on 24-25 February 2015 in Paris.

OECD (2013), *Water and Climate Change Adaptation: Policies to Navigate Uncharted Waters*, OECD Studies on Water, OECD Publishing, Paris, *http://dx.doi.org/10.1787/9789264200449-en.*

OECD (2012a), *OECD Environmental Outlook to 2050: The Consequences of Inaction*, OECD Publishing, Paris, *http://dx.doi.org/10.1787/9789264122246-en.*

OECD (2012b), "Meta-analysis of stated preference VSL studies: Further model sensitivity and benefit transfer issues", Working Party on National Environmental Policies, ENV/EPOC/WPNEP(2010)10/FINAL, OECD Publishing, Paris.

OECD (2011), *How's Life?: Measuring Well-being*, OECD Publishing, Paris, *http://dx.doi.org/10.1787/9789264121164-en.*

OECD (2001), *Multifunctionality: Towards an Analytical Framework*, OECD Publishing, Paris, *http://dx.doi.org/10.1787/9789264192171-en.*

Pendleton, L. et al. (2009), "Estimating the Potential Economic Impacts of Climate Change on Southern California Beaches", draft paper prepared for the California Energy Commission and the California Environmental Protection Agency, *www.energy.ca.gov/2009publications/CEC-500-2009-033/CEC-500-2009-033-D.PDF.*

Pickering, T.D. et al. (2011), "Vulnerability of aquaculture in the tropical Pacific to climate change", in *Vulnerability of Tropical Pacific Fisheries and Aquaculture to Climate Change* [Bell, J.D., J.E. Johnson and A.J. Hobday (eds.)]. Secretariat of the Pacific Community, Noumea, New Caledonia, pp. 647-731.

Pindyck, R.S. (2013), "Climate change policy: What do the models tell us?", *Journal of Economic Literature*, Vol. 51(3), pp. 860-872.

Pindyck, R.S. (2012), "Uncertain outcomes and climate change policy", *Journal of Environmental Economics and Management*, Vol. 63, pp. 289-303.

Schewe, J. et al. (2013), "Multimodel assessment of water scarcity under climate change", *Proceedings of the National Academy of Sciences (PNAS)*, Vol. 111 (9), pp. 3245-50.

Seo, S.N. and R. Mendelsohn (2008), "Animal husbandry in Africa: climate change impacts and adaptations", *African Journal of Agricultural and Resource Economics*, Vol. 2, pp. 65-82.

Shamsuddoha, M. and R.K. Chowdhury (2009), "Climate Change Induced Forced Migrants: In need of dignified recognition under a new Protocol", Bangladesh: Campaign of Equity and Justice Working Group Bangladesh (Equitybd), available at: *www.glogov.org/images/doc/equitybd.pdf* (10 March 2015).

Shogren, J.F. (1997), "Economics and the Endangered Species Act", University of Michigan, available at: *www.umich.edu/~esupdate/library/97.01-02/shogren.html* (18 March 2015).

Staddon, P.L., H.E. Montgomery and M.H. Depledge (2014), "Climate warming will not decrease winter mortality", *Nature Climate Change*, Vol. 4 (March 2014), pp. 190-194.

Stanke C., V. Murray, R. Amlôt, J. Nurse and R. Williams (2012), "The Effects of Flooding on Mental Health: Outcomes and Recommendations from a Review of the Literature", *PLOS Currents Disasters*, 2012 May 30, Ed. 1, doi: *10.1371/4f9f1fa9c3cae.*

Steininger, K.W. et al. (2015), *Economic Evaluation of Climate Change Impacts: Development of a Cross-Sectoral Framework and Results for Austria*. Springer International Publishing, Switzerland.

Stern, N. (2013), "The structure of economic modelling of the potential impacts of climate change: Grafting gross underestimation of risk onto already narrow science models", *Journal of Economic Literature*, Vol. 51(3), pp. 838-859.

Stern, N. (2007), *Stern Review: The Economics of Climate Change*, Cambridge: CUP.

Stiglitz, J.E., A. Sen and J.-P. Fitoussi (2010), "Report by the Commission on the Measurement of Economic Performance and Social Progress", *www.stiglitz-sen-fitoussi.fr/documents/rapport_anglais.pdf.*

Strzepek, K. et al. (2014), "Benefits of greenhouse gas mitigation on the supply, management, and use of water resources in the United States", *Climatic Change*, November 2014.

Sue Wing, I. and E. Lanzi (2014), "Integrated assessment of climate change impacts: Conceptual frameworks, modelling approaches and research needs", *OECD Environment Working Papers*, No. 66, OECD Publishing, Paris, *http://dx.doi.org/10.1787/5jz2qcjsrvzx-en.*

Takahashi, K., Y. Honda and S. Emori (2007), "Assessing mortality risk from heat stress due to global warming", *Journal of Risk Research*, Vol. 10(3), pp. 339-354.

Tacoli, C. (2009), "Crisis or adaption? Migration and climate change in a context of high mobility", *Environment and Urbanization*, Vol. 21 (2), pp. 513-525.

TEEB (2014), "Ecosystem Services", website article of The Economics of Ecosystems and Biodiversity (TEEB), *www.teebweb.org/resources/ecosystem-services/* (accessed 15 March 2015).

UK Treasury (2003), *The Green Book: Appraisal and evaluation in central government*, London.

UN (2001), "Announcing the release of the World Atlas of Coral Reefs", press release, United Nations Coral Reef Unit, *http://coral.unep.ch/atlaspr.htm* (accessed 16 March 2015).

UNDP (2014), *Human Development Report, Sustaining Human Progress: Reducing Vulnerabilities and Building Resilience*, Washington, DC, USA.

UNDP (1990), *Human Development Report*, Oxford University Press, New York/Oxford.

USGCRP (2009), *Global Climate Change Impacts in the United States*, United States Global Change Research Program (Karl, T.R., J.M. Melillo and T.C. Peterson [eds.]), Cambridge University Press, New York, NY, USA.

Waldinger, M. (2015), "The effects of climate change on internal and international migration: Implications for developing countries", *Grantham Research Institute on Climate Change and the Environment Working Paper* No. 12, London School of Economics, London.

Wall, E., A. Wreford, K. Topp and D. Moran (2010), "Biological and economic consequences heat stress due to a changing climate on UK livestock", *Advances in Animal Biosciences*, Vol. 1(1), pp. 53.

Ward, P.J. et al. (2014), "Strong influence of El Niño Southern Oscillation on flood risk around the world", *Proceedings of the National Academy of Sciences*, Vol. 111(44), pp. 15659-15664.

Ward, P.J. et al. (2013), "Assessing flood risk at the global scale: Model set up, results, and sensitivity", *Environmental Research Letters*, Vol. 8, 044019.

Warren, R., C. Hope, M. Mastrandrea, R.S.J. Tol, W.N. Adger and I. Lorenzoni (2006), "Spotlighting impacts functions in integrated assessment", *Tyndall Centre Working Papers*, No. 91, Tyndall Centre for Climate Change Research.

Watkiss, P. and A. Hunt (2012), "Projection of economic impacts of climate change in sectors of Europe based on bottom up analysis: human health", *Climatic Change*, Vol. 112, pp. 101-126.

Weitzman, M.L. (2013), "Tail-hedge discounting and the social cost of carbon", *Journal of Economic Literature*, Vol. 51(3), pp. 873-882.

Weitzman, M.L. (2012), "GHG targets as insurance against catastrophic climate damages", *Journal of Public Economic Theory*, Vol. 14(2), pp. 221-244.

Weitzman, M.L. (2009), "On modeling and interpreting the economics of catastrophic climate change", *The Review of Economics and Statistics* Vol. 91(1), pp. 1-19.

WHO (2014), *Quantitative Risk Assessment of the Effects of Climate Change on Selected causes of Death, 2030s and 2050s*, World Health Organization, Geneva.

Wilkinson, P. et al. (2007), "A global perspective on energy: Health effects and injustices", *Lancet*, Vol. 370(9591), pp. 965-978.

Winsemius, H.C. and P.J. Ward (2015), "Projections of future urban damages from floods", personal communication.

World Bank (2011), "Study on impacts and costs of forced displacement. Volume II: State of the art literature review", *http://siteresources.worldbank.org/EXTSOCIALDEVELOPMENT/Resources/244362-1265299949041/6766328-1265299960363/VolumeIILiterature.pdf*.

World Bank (2012a), *Guidelines for Assessing the Impacts and Costs of Forced Displacement*, Washington, DC.

World Bank (2012b), "Assessing the Impacts and Costs of Forced Displacement. Volume I: A Mixed Methods Approach", *http://siteresources.worldbank.org/EXTSOCIALDEVELOPMENT/Resources/244362-1265299949041/6766328-1265299960363/ImpactCostStudy-VolumeI2.pdf*.

WRI (2014), "Aqueduct Water Risk Atlas", World Resources Institute, *www.wri.org/our-work/project/aqueduct/aqueduct-atlas*.

Chapter 4

The benefits of policy action

This chapter opens with a discussion on how policy makers can deal with climate change even if future damages and other consequences are not fully known. Then, stylised projections of the AD-DICE model are used to highlight the potential benefits of adaptation and mitigation policies at the global level, and the trade-offs and synergies between both policy options. The chapter closes with an assessment of the avoided damages from mitigation policy action in the ENV-Linkages model, and compares the sectoral distribution of damages and mitigation costs.

The consequences of climate change, as outlined in detail in the previous chapters, with losses in gross domestic product (GDP) for almost all regions and numerous important other consequences, imply a strong call for policy action. By implementing ambitious mitigation policies to reduce the emission sources of climate change, and adaptation policies to best deal with the remaining consequences, the worst impacts may be avoided, and damages from climate change substantially reduced. These benefits of policy action can be compared to the costs of the policies to assess the optimal level of government intervention.

This report does not aim to provide a definite answer to the question of optimal policies for climate change, nor does it aim to have the breadth of IPCC reports. Rather, by recognising the complexity of the framework that is needed for an adequate assessment, and the lack of reliable quantitative information on a number of essential aspects, it limits itself to the much narrower issue of how – within the context of the modelling framework presented in Chapter 1 – the costs of inaction as discussed in Chapters 2 and 3 may be reduced by adaptation and mitigation policies. This includes insights on sectoral bottlenecks, i.e. specific sectors that may be negatively affected by climate change as well as by mitigation policies. A detailed cost-benefit analysis of the regional and sectoral consequences of specific policies is left for future study.

4.1. Policy making under uncertainty for inter-temporal issues

The projections presented in Chapters 2 and 3 are surrounded by large uncertainties. For instance, the alternative damage specifications based on the suggestions by Weitzman, as presented in Figure 3.2, fall well outside the uncertainty range around the central projection stemming from different assumptions regarding the climate sensitivity parameter. This re-enforces the notion that the uncertainty ranges given throughout this paper only reflect one particular source of uncertainty – albeit an important one – concerning the equilibrium climate response to a doubling of carbon concentrations (climate sensitivity). Other uncertainties, such as impacts of catastrophic events and uncertainties on the socioeconomic drivers of growth, should also be taken into account, but cannot easily be quantified and are therefore omitted from the uncertainty ranges presented in this report.

A single central projection of global climate damages is insufficient to portray a robust message on the links between climate change and economic growth. Further exploration of the uncertainties could include: i) the formulation of different scenarios for baseline projections, reflecting uncertainty around some key drivers such as demography, the long-term trend of economic growth and natural resources availability; ii) further investigating the role of adaptation as a means to limit negative impacts and boost positive ones; iii) comparing different representations of the climate system, either through the use of different underlying climate models (as suggested by e.g. Warszawski et al., 2014) which would also help to shed light on the uncertainties in the regional patterns of climate change, or – as a minimum – by varying the climate sensitivity, as done in this report in a stylised manner. Consequently, this report needs to be seen as the assessment of one possible projection of the economic consequences of climate change. A more robust, comprehensive assessment would rely on comparing different models and studies, but would require significant research investments.

Policies designed to reduce the costs of climate damage need to be designed to be robust in the face of uncertainties (see Box 4.1).[1] The Intergovernmental Panel on Climate Change (IPCC, 2014a) suggests that "the social benefits from investments in mitigation tend to increase when uncertainty in the factors relating greenhouse gas (GHG) emissions to climate change impacts are considered (medium confidence)."

Box 4.1. **Policy making under risk and uncertainty**

The OECD has a long strand of work on robust policy making under risk and uncertainty which continuously underlines that "you can't manage what you can't measure". Building on earlier work on global shocks (OECD, 2010), OECD (2014d) presents a recommendation on the governance of critical risks, recommending that countries "establish and promote a comprehensive, all-hazards and transboundary approach to country risk governance to serve as the foundation for enhancing national resilience and responsiveness". This incorporates supporting a comprehensive approach to critical risks, ensuring preparedness, raising awareness, building an adaptive capacity in crisis management, all to be done in a transparent and accountable manner. Applications to specific policy domains, e.g. on water security (OECD, 2013), stress the use of risk-based approaches and outline the key success factors: know the risk; target the risk; and manage the risk (OECD, 2013). The risk approach is also highlighted in OECD work on adaptation to climate change (OECD, 2015c).

The OECD (2010) also stresses that policy makers can (and have to) deal with "previously unknown hazards for which there are no data and no model for likelihood and impacts":

"Managing unknown-unknowns might seem like guesswork, but there are several strategic concepts available to aid risk managers. Generally, this involves a combination of two techniques:

1. Designing or reinforcing complex systems to be more robust, redundant and/or diverse as appropriate; and
2. Building societal resilience to unknown events by drawing from experience with extreme events that share some similarity in nature or scale." (OECD, 2010).

Policy making under uncertainty goes beyond preparing for risks to the economic system, and also covers the need for flexible policy frameworks that can adapt to new information and identifying no-regret options that are good for economic growth regardless of the future state of the economy and environment. Robust policy making also accounts for the (uncertain) benefits and costs of policies that cannot be captured easily in cost-benefit analysis and can under some circumstances be linked to the precautionary principle. For instance, option values and risk premia can be used in cost-benefit analysis to select those policies that reduce uncertainty over ones that have riskier outcomes. One example where robust policy regimes are discussed is in the context of water allocation rules (OECD, 2015a).

The need to incorporate long-term systemic threats into the core tools for government support is also highlighted in Braconier et al. (2014). That report, as well as the current analysis, are part of the OECD-wide strategy on New Approaches to Economic Challenges (OECD, 2015b), which stresses the need for an open perspective on potential risks and uncertainties, and broadening the use of scenario analysis to support policy making.

Identifying appropriate policies for mitigation and for adapting to remaining climate damages also requires a long-term perspective. Despite short-term policy costs, climate change policies have significant benefits. First, policies that achieve emission reductions in the near term deliver a stream of future benefits by reducing climate change impacts, while adaptation helps reduce the adverse consequences of climate impacts that are already underway. Secondly, there are important co-benefits from policy action that can be reaped immediately. As highlighted in Chapter 3, flood risks, the welfare costs of premature deaths and the increased risks for large-scale singular events are of particular importance. Co-benefits outside the climate domain include improved health from reduced air pollution (see e.g. Nemet et al., 2010; Bollen and Brink, 2014). These co-benefits are potentially very important but could not be incorporated in the numerical assessment provided by this report. [2]

4.2. Economic trade-offs between adaptation, mitigation and climate damages

As explained in Chapter 1, the ENV-Linkages model cannot be used to project the most economically efficient combination of adaptation, mitigation and damages over time. Hence, the strategy here is to use the more stylised integrated assessment model AD-DICE, which focuses on these inter-temporal aspects, to provide insights into the costs and benefits of policy action on climate change. The strength of the AD-DICE model is that it can assess both mitigation and adaptation policies, as well as investigate their interactions (De Bruin et al., 2009a; 2009b; Agrawala et al., 2011).

Adaptation policies are essential to keep the costs of climate change impacts as low as possible. These include direct government intervention where necessary, e.g. for large-scale infrastructure projects, as well as facilitating market-driven adaptation by private actors, e.g. to overcome information barriers and moral hazard issues. In Section 4.2.1, the AD-DICE model is used to investigate stylised adaptation policy scenarios, specifically looking at how different levels of adaptation affect the costs of climate change impacts and the costs of implementing the adaptation measures. Given the wide variety of possible adaptation measures, and the fact that most of them are at the local scale, these adaptation scenarios are necessarily stylised, and focus on two specific scenarios, namely "optimal adaptation" and "no adaptation".

The effectiveness of adaptation notwithstanding, mitigation policies are needed to limit climate change and thus avoid much of the damages especially in the long run, limit the risks and avoid tipping points. Avoiding the long-term consequences of emissions requires immediate policy action. Justification for such mitigation actions cannot be based directly on the time profile of damages as they arise. Ideally, they should be based on the full stream of future avoided (market) damages stemming from current emission reductions, plus a premium to manage the risks of non-market damages, catastrophic events and crossing irreversible tipping points. In Section 4.2.2, the AD-DICE model is used to identify least-cost mitigation pathways by maximising the net present value of the full pathway of avoided damages, taking the costs of emission reductions into account. Interactions between adaptation and mitigation policies are analysed in Section 4.2.3. The inter-temporal aspect that is at the core of climate change policies necessitates the use of a discount rate to compare GDP impacts and avoided damages over the whole time frame. This section therefore also presents a sensitivity analysis on the discount rate assumptions.

AD-DICE does not include specific policy options for mitigation or adaptation, but rather synthesizes these options into smooth functions for the costs and benefits of reducing emissions (mitigation) or remaining damages (adaptation): effectively, this entails economy-wide carbon taxation as mitigation policy, and enabling stock adaptation as adaptation policy. Reducing emissions and damages by a little amount will deliver a relatively large gain at relatively low costs, but as the policies become more stringent, the incremental costs increase while the incremental gains become smaller. By comparing these incremental costs with the incremental benefits, an "optimal" policy level is assessed. At the optimum, emissions and damages are reduced until the incremental costs equal the incremental gains, and no further welfare improvement from further strengthening the policies can be achieved. At this point, implicitly all "least-cost" options, i.e. options that lead to net welfare gains, are adopted, while more costly options are not. In this evaluation of incremental costs and gains, the entire pathway of consumption is considered to maximise the net present value of utility.

4.2.1. *Benefits of adaptation policy*

AD-DICE, adapted from the DICE model (Nordhaus and Boyer 2000, Nordhaus 2012), was the first and is still one of the few integrated assessment models (IAMs) that explicitly model adaptation as a macroeconomic policy variable.[3] Thus, the model can be used to investigate how the damages from climate change are affected by different adaptation policy scenarios (cf. Agrawala et al., 2011). Adaptation can take on two forms in the AD-DICE model, namely stock and flow. Stock adaptation refers to adaptation measures that require investments beforehand to build adaptation capital. This adaptation stock reduces damages of climate change in the future. Flow adaptation refers to adaptation measures that do not require investments beforehand but where benefits are reaped almost instantaneously. Government involvement can facilitate the efficient application of this adaptation (by e.g. overcoming knowledge barriers), but is not necessary for its implementation.

The literature identifies many restrictions to adaptation and, without appropriate government policies, adaptation is expected to fall short of the societal optimal amount (UNEP 2014). Assuming that all adaptation options are readily available to firms and households and will be implemented without government intervention will hence result in lower residual damage estimates than are likely to occur, specifically without targeted adaptation policies. In principle, the central projection should include all adaptation efforts that are driven by market forces, whereas actions that require government intervention should be excluded. This distinction is not available in AD-DICE, but as a (necessarily crude) approximation and following De Bruin (2014) and UNEP (2014), stock adaptation is considered to be public and not market driven and would hence need government coordination for its successful implementation. Flow adaptation is assumed to be private and market driven. These assumptions are in line with the way stock and flow adaptation have been calibrated in the AD-DICE model (De Bruin, 2014). Thus, the central projection reflects the "no adaptation policies" situation and includes flow adaptation, while stock adaptation investments are excluded. This approximation aligns with earlier assessments of the need for government intervention in adaptation (e.g. Agrawala et al., 2011).

In the "Full Adaptation" scenario, the assumption is made that adaptation levels are chosen to minimise costs ("optimal adaptation", not referring to avoiding all possible damages but rather to the implementation of all least-cost stock and flow adaptation options). As shown in Figure 4.1, the least-cost level of adaptation – measured as the percentage of potential damages (gross damages) in a given period that is avoided through past and current adaptation – rises over time.[4] When including investments in adaptation stock (going beyond the flow adaptation only assumption in the central projection), the

Figure 4.1. **Percentage of damages from selected climate change impacts addressed by adaptation**

Percentage reduction of potential (gross) damages

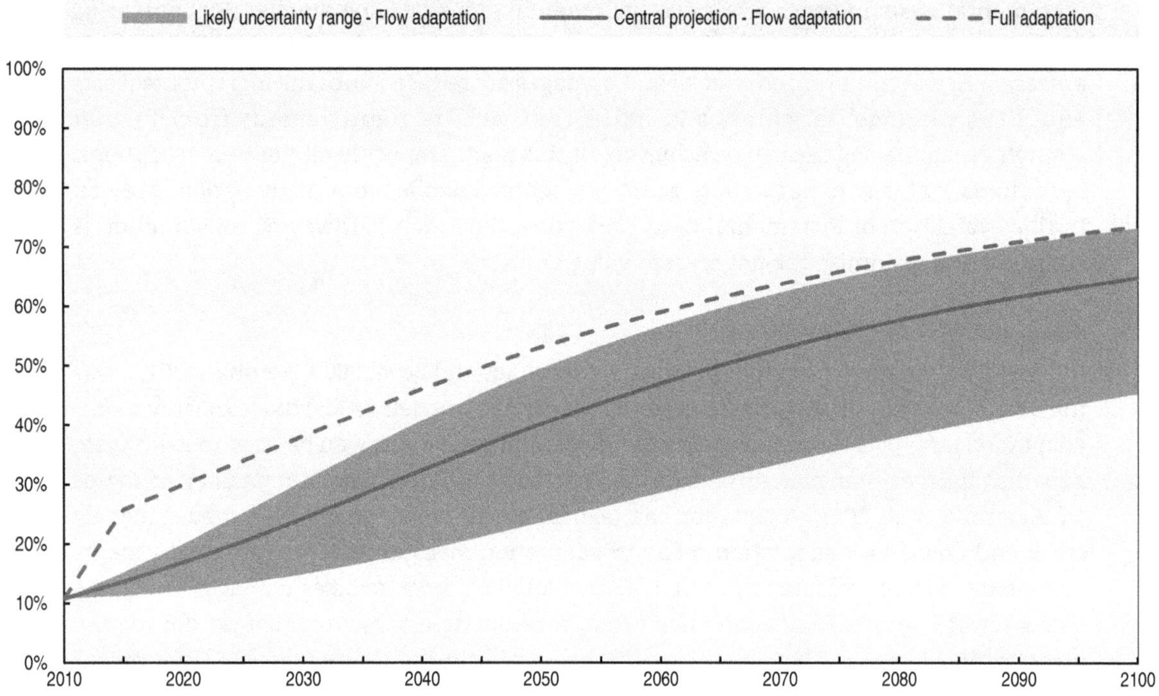

Source: AD-DICE model.

StatLink http://dx.doi.org/10.1787/888933276127

adaptation levels rise much faster, highlighting that both types of adaptation are important to keep damages low and that it pays off to immediately start investing in the adaptation stock.

In the hypothetical "No adaptation" scenario, no further investments in adaptation stock are made nor does flow adaptation occur. In this extreme case, regions determine their optimal mitigation and consumption levels given the gross damages of climate change without the possibility of reducing gross damages through adaptation. Hence the existing adaptation stock provides a small amount of avoided damages in the short run, but the adaptation stock quickly depreciates. As in this scenario flow adaptation is also excluded, the adaptation level diminishes to zero. The motivation for investigating this scenario is not its realism, but rather to show the potential size of damages when there are severe barriers to market adaptation actions.

In the absence of mitigation (i.e. the central projection as discussed in Chapter 3), damages increase more than proportionally over time, reaching almost 6% of GDP by 2100 (with a likely uncertainty range on the equilibrium climate sensitivity, ECS, of 2 to 10%). Again, it should be stressed that this represents just one source of uncertainty, while there are many others surrounding these projections (see Chapter 2).

Not adapting to climate change impacts at all tends to increase the damages especially in the longer run (with damages in the central projection amounting to 11% of annual GDP by the end of the century), cf. Figure 4.2. The uncertainty range of the GDP impacts (due to uncertainty on the equilibrium climate sensitivity only) is also substantially larger in absence of adaptation. Combining the assumption of no adaptation with high climate sensitivity (likely range) leads to a projected upper bound on potential GDP impacts that

Figure 4.2. **Climate change damages from selected climate change impacts for different adaptation scenarios**

Percentage change w.r.t. no-damage baseline

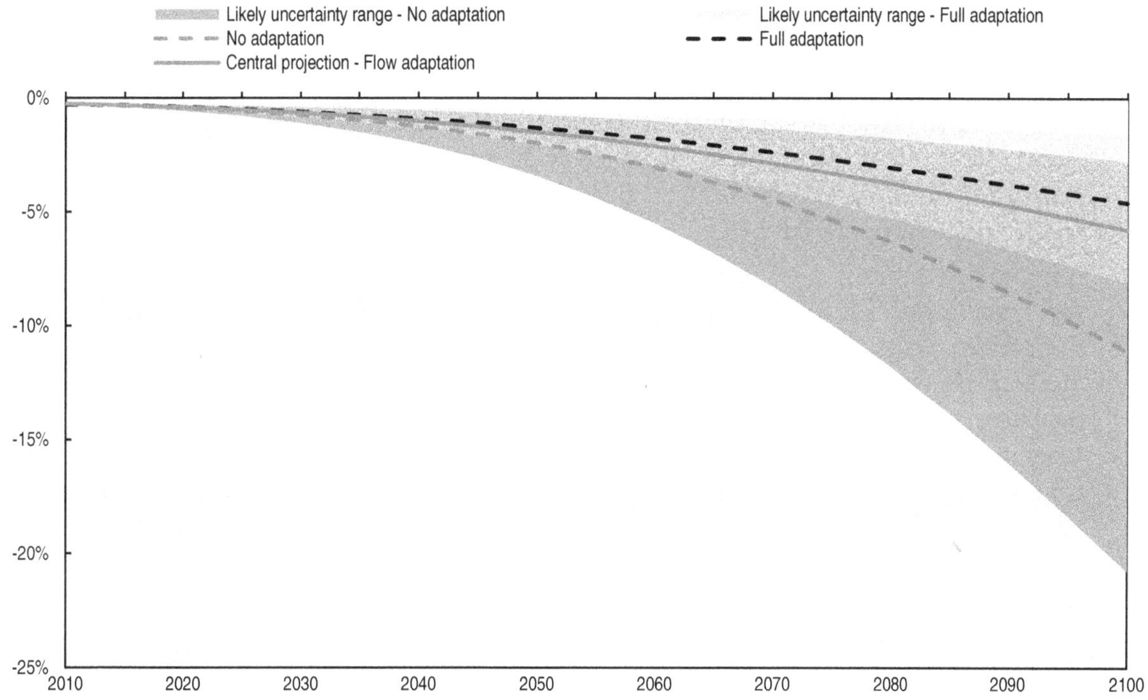

Source: AD-DICE model.

StatLink http://dx.doi.org/10.1787/888933276138

reach double digits well before the end of the century.[5] Although the assumption of no adaptation is extreme (when faced with the consequences of climate change, households and firms will adapt their behaviour, even in absence of government policies), this shows that it is vital to facilitate market adaptation actions, and not hamper them.

The "Full adaptation" scenario includes all least-cost measures to adapt to climate change, whether they are pro-active, stock-type measures or reactive, flow-type. When not only the responsive, flow adaptation measures are available (as in the central projection), but also stock measures are implemented, damages are still increasing over time, emphasising that adaptation cannot take away all impacts from climate change. What is clear from the uncertainty ranges, however, is that adaptation is a powerful instrument to avoid the largest gradual damages from climate change (though not catastrophic events), and greatly limit the risks (even though the benefits of adaptation are in themselves also uncertain). The benefits of adaptation policies alone amount to more than 1 per cent-point of GDP by the end of the century (and the corresponding benefits of both market- and policy-driven adaptation amount to more than 6 percentage-points of GDP). Of course, these benefits are surrounded by the same uncertainties as the costs of inaction.

4.2.2. *Benefits of mitigation policy*

The optimal level of emission reductions results from equating the marginal costs of one unit of additional emission reduction with the discounted stream of additional avoided damages, i.e. the marginal benefits of a unit of emission reduction. Therefore, these projections depend crucially on the discount rate that is used in the analysis. The graphs presented here are based on the discount rates as proposed by the UK Treasury (2003); the resulting discount factors and comparison with alternative assumptions are investigated in Section 4.2.4 below. Although many pathways are feasible to achieve a long-term climate target, such as a 450 ppm stabilisation scenario, it pays to start reducing emissions immediately, despite relatively low damage levels, to take advantage of existing cheap reduction options and avoid the need for very rapid increases in mitigation levels later (cf. OECD, 2012).[6] It is important to remember that these mitigation pathways are based only on the benefits that are included in the model, and exclude a risk premium due to large-scale singular events (Section 3.2.8) and co-benefits from improved air quality.

Figure 4.3 shows how in a least-cost scenario ("optimal mitigation") emission reduction rates, i.e. the percentage reduction of emissions below the no-damage baseline projection, are projected to evolve over time, assuming flow adaptation. Even in the absence of (new) mitigation policies, some emission reductions will occur due to the depletion of fossil fuels and the already implemented mitigation policies around the world; together, these imply roughly 10% emission reductions, and this is slowly increasing over time as fossil fuels become scarcer. The least-cost mitigation pathway as projected with AD-DICE implies an immediate jump in the mitigation rate and a stable increase afterwards. Compared to the no policy action scenario, emissions are immediately reduced below 2010 levels. The lower the equilibrium climate sensitivity, the lower the benefits from emission reduction are, and hence the lower the mitigation rates and the larger the flexibility to adjust the timing of emission reductions in the least-cost emission pathway.

Figure 4.3. **Global emission reduction rates for least-cost mitigation scenario**

Percentage of no-damage baseline

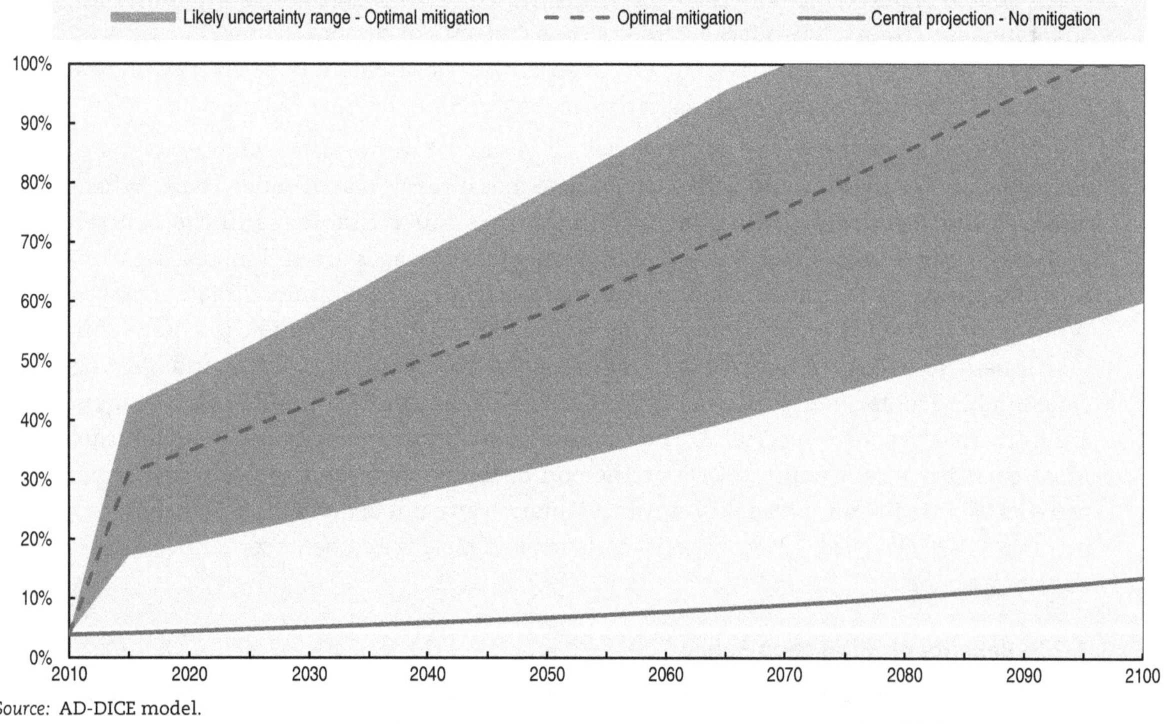

Source: AD-DICE model.

StatLink http://dx.doi.org/10.1787/888933276143

Mitigation efforts reduce the impacts of climate change but do not completely take them away. It is too costly to aim to prevent all damages from climate change beyond current levels (and historical emissions have already committed society to some damages; cf. Chapter 3). Figure 4.4 shows how least-cost (or optimal) mitigation efforts reduce damages to around 2.5% of GDP. Thus, the benefits of mitigation action rapidly increase over time to more than 3 per cent-points of GDP by the end of the century.

Figure 4.4. **Climate change damages from selected climate change impacts for different mitigation scenarios**

Percentage change w.r.t. no-damage baseline

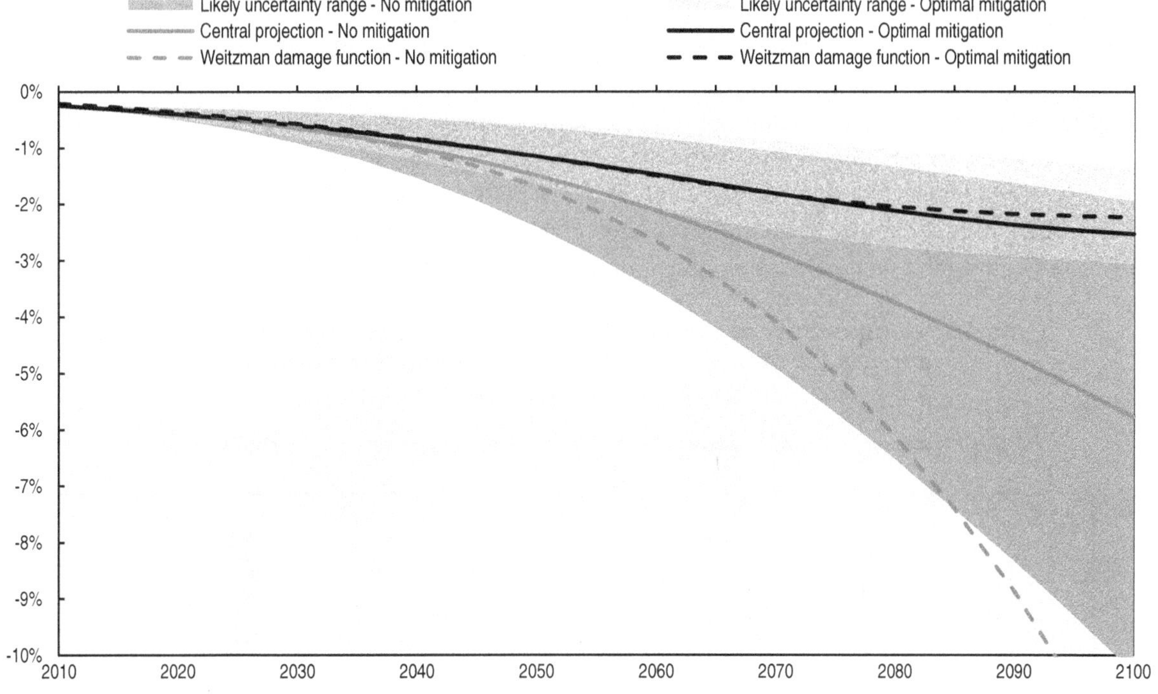

Source: AD-DICE model.

StatLink http://dx.doi.org/10.1787/888933276156

The likely uncertainty range on the damage level, reflecting the equilibrium climate sensitivity also becomes much smaller. As carbon concentrations are lower in the mitigation scenario, the corresponding range in temperature increases scale down more than proportionately, leading in turn to a more than proportional reduction in damages. Consequently, by 2100, the likely uncertainty range of damages for the least-cost mitigation policy is 1%-3%, whereas it is 2%-10% for the "No mitigation" scenario.

The Weitzman damage function (Weitzman, 2012) leads to a starker reduction in damages from mitigation than the central projection (compare the dashed and solid line in Figure 4.4). The logic is that the main difference between the Nordhaus damage function and Weitzman's alternative is in the level of damages when temperature increases go beyond 2-3 degrees; for temperature increases below 2 degrees, both functions are by construction very similar. In the least-cost mitigation scenarios, temperature increases remain smaller and hence there is hardly any difference between the Nordhaus and Weitzman damage functions.[7] However, this comes at a high total climate cost in the Weitzman specification as mitigation costs are higher throughout the century in this alternative.

4.2.3. *Interactions between adaptation and mitigation policies*

Section 4.2.1 highlighted that in the absence of mitigation action, adaptation measures play an important role in limiting the damages from climate change. Figure 4.5, which presents the different adaptation costs, mitigation costs and damages for the middle and the end of the century, shows that the role of adaptation is much smaller when damages are limited by least-cost mitigation action. Adaptation and mitigation are both powerful instruments to limit climate damages, especially when the other instrument is lacking. In terms of cost minimisation, however, both policies partially substitute each other.[8] The least-cost policy package will consist of both measures, but Figure 4.5 clearly shows that adaptation cannot be a perfect substitute for mitigation. If only adaptation policies are available ("Full Adaptation – No mitigation" scenario), damages are substantially larger than when only mitigation policies are available ("No adaptation – Optimal mitigation" scenario), especially in the first half of the century. By the end of the century, the total climate change costs are roughly equal when optimal mitigation policies are available, regardless of the availability of adaptation. But even with optimal mitigation, adaptation can significantly reduce the costs of climate change earlier in the century; the cumulative costs of climate change over the century (not shown in the figure) are 17% higher when adaptation policies are not available (and 66% higher when market-driven adaptation is also not available).

Figure 4.5. **Components of climate change costs from selected climate change impacts for different adaptation and mitigation scenarios**

Percentage of no-damage baseline GDP level

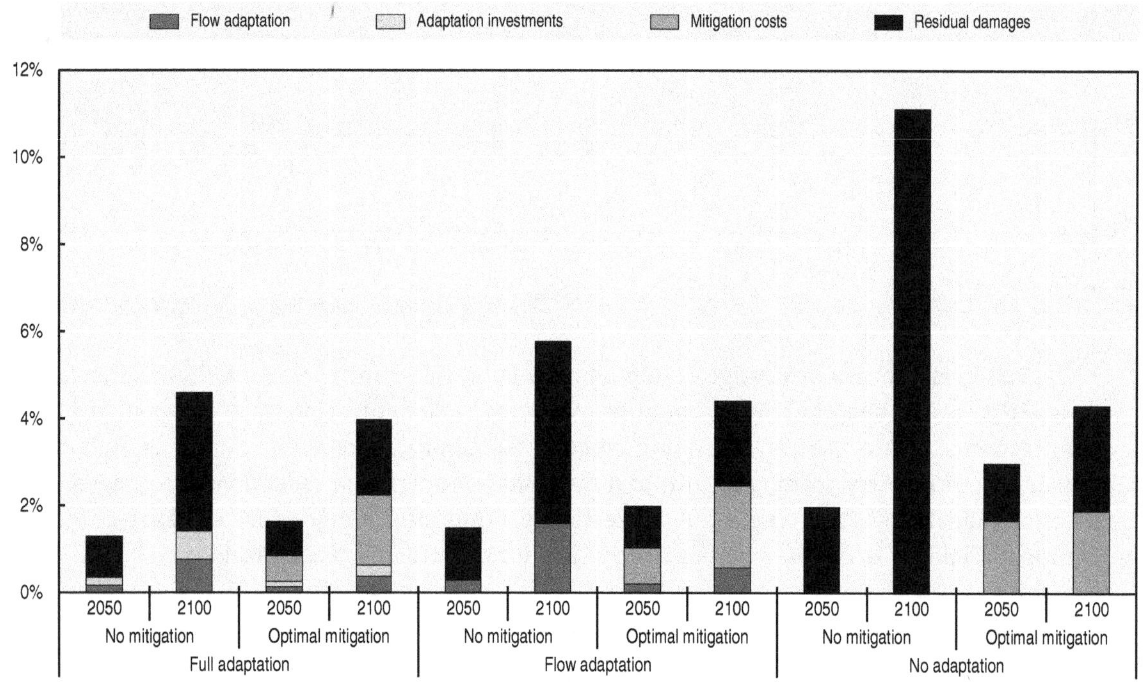

Flow adaptation | Adaptation investments | Mitigation costs | Residual damages

12% 10% 8% 6% 4% 2% 0%

2050 | 2100 | 2050 | 2100 | 2050 | 2100 | 2050 | 2100 | 2050 | 2100 | 2050 | 2100

No mitigation | Optimal mitigation | No mitigation | Optimal mitigation | No mitigation | Optimal mitigation

Full adaptation | Flow adaptation | No adaptation

Source: AD-DICE model.

StatLink http://dx.doi.org/10.1787/888933276168

4.2.4. *Sensitivity analysis on the discount rate*

The speed and rate of emission reductions depends among other things on the value placed on future damages. With lower discount rates, relatively more emphasis is placed on future costs and benefits, and as the costs of emission reductions precede most of the benefits (costs now lead to a stream of future avoided damages), lower discount rates imply more stringent mitigation policies in the near term. Hence, the discount rate plays an important role in the projection of the least-cost mitigation pathway.

Most economic models use the so-called Ramsey rule to calculate the financial discount rate. This rule uses a "pure rate of time preference" (PRTP), an inter-temporal income inequality aversion elasticity and the growth rate of GDP (Ramsey, 1928; Nordhaus, 2007; Gollier, 2007). In the original DICE and RICE models, a PRTP of 1.5% and an elasticity value of 2 are chosen (Nordhaus, 2011); but in the latest DICE model Nordhaus (2012) has revised his elasticity value to 1.45 (which is used in the analysis here). In the Stern report, PRTP is chosen to be 0.1% and the elasticity 1 (Stern, 2007). An intermediate case has been proposed by the UK Treasury, with a PRTP of 1.5% and an elasticity of 1 (UK Treasury, 2003). The net present value of future policy action is calculated as the sum of future benefits minus future costs, each discounted back to the current period using the time-specific discount rates. Three helpful indicators can shed light on the influence of the discount rate: i) the annual discount rate in the first decade, ii) the present value in 2010 of 100 USD in 2060, and iii) the share of the infinite horizon that is captured after 50 years. These indicators are given in Table 4.1 for the different Ramsey discount rate rules discussed above (including those for the RICE2010 model, although these are not further used in the analysis).

Table 4.1. **Effects of different discount rates**

	UK Treasury discounting	Nordhaus discounting DICE2013 (RICE2010)	Stern discounting
Annual discount rate first decade	4.2%	5.2% (6.6%)	2.5%
Present value of receiving USD100 in 2060	16.7	10.7 (6.0)	34.1
Share of first 50 years in infinite horizon	79%	86% (92%)	53%

Source: Own calculations using the AD-DICE model.

StatLink http://dx.doi.org/10.1787/888933276294

In the absence of mitigation policies, the discount rate does not have a major influence on the level of damages in percentage of GDP. The only difference is through an adjustment of the investments in stock adaptation, but this effect is relatively minor. The influence on the least-cost pathway is, however, much larger, as shown in Figure 4.6. Essentially, the lower the discount rate, the lower the least-cost damage levels are. This is because the present value of the future stream of avoided damages is larger and mitigation becomes more cost-effective, increasing mitigation and hence decreasing climate change and its concomitant damages. A second result is that the likely uncertainty range is smaller when lower discount rates are used. This follows the same mechanism as explained above for the reduction in uncertainty range when moving from no mitigation to least-cost mitigation levels.

Finally, as the discount rate affects the weighing of costs and benefits over time, it also influences the pathway of global average temperature increases that would be experienced if least-cost mitigation is implemented, as illustrated in Figure 4.7. For the Nordhaus discount rate, with a large weight on short-term costs, it is optimal to let temperatures increase to 3 degrees above pre-industrial levels before the end of the century. Using the UK Treasury discount rate values (as in the central projection), the temperature increase is more modest and remains well below 3 degrees, while Stern discount rates limit temperature increases even further to below 2 degrees. As illustrated in the figure, these reflect just a probability that temperatures remain at those levels; there is a substantial chance that temperatures are higher, when the equilibrium climate sensitivity is higher.

Figure 4.6. **Climate change damages from selected climate change impacts for the least-cost mitigation scenario**

Percentage change w.r.t. no-damage baseline

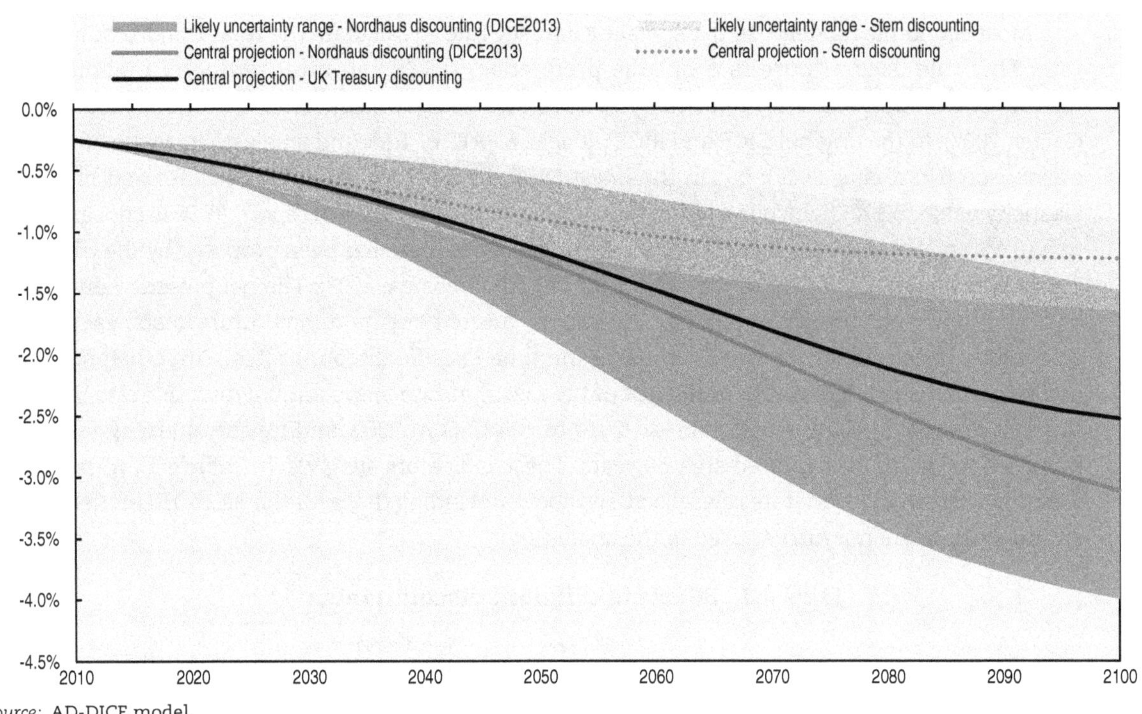

Source: AD-DICE model.

StatLink http://dx.doi.org/10.1787/888933276176

Figure 4.7. **Temperature increases with optimal mitigation, different discount rates**

Degrees Celsius above pre-industrial level

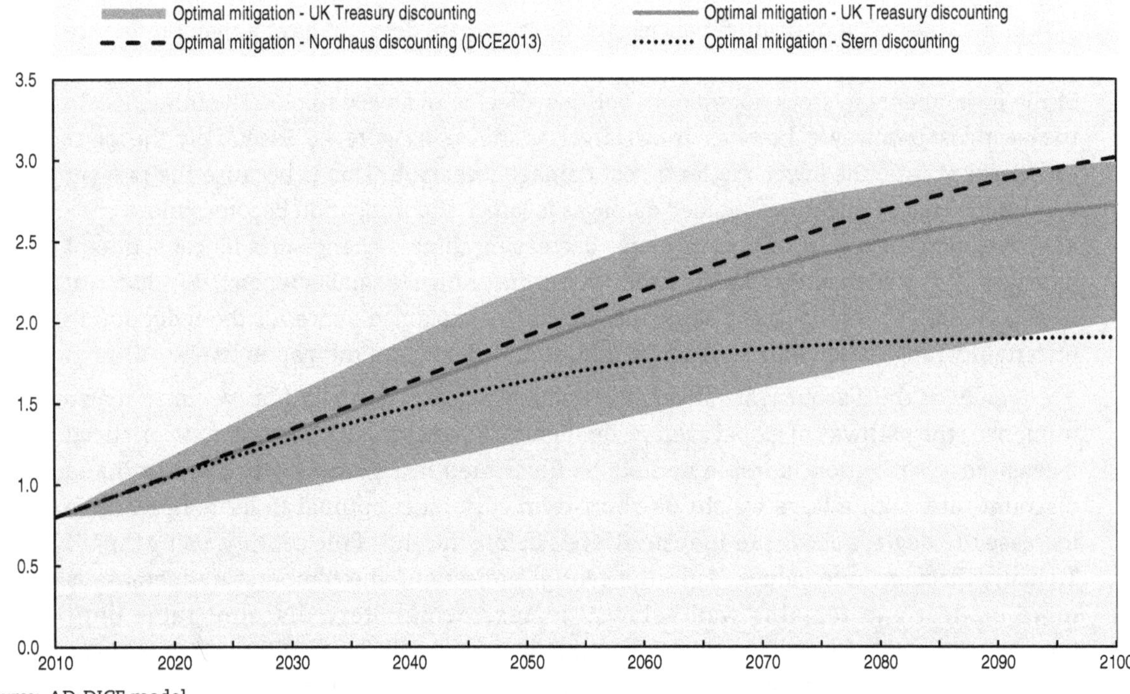

Source: AD-DICE model.

StatLink http://dx.doi.org/10.1787/888933276184

4.3. Sectoral and regional consequences of mitigation action

As mentioned above, the ENV-Linkages model is not suited for a direct cost-benefit analysis of various climate policies. However, ENV-Linkages has better capability to simulate sectoral and regional economic activity, and provides more detailed insights into the different damage categories. Furthermore, in ENV-Linkages, carbon pricing policies can be used to shed light on the avoided damages of mitigation policy action, i.e. the change in the level of damages from current and past mitigation actions. It can also compare the regional and sectoral patterns of mitigation costs with the costs of inaction (damages). In line with the central projection for optimal mitigation policy in AD-RICE, the ENV-Linkages model is used to simulate the emission reduction pathway that follows the pathway for optimal mitigation in the central projection for the UK Treasury discounting scheme in Section 4.2.2. The policy is implemented through a hypothetical global carbon market. This purely serves as a reference point, as it ensures that the least-cost reduction options are implemented in the model, without additional costs associated with specific instrument choices. Given that regional mitigation costs depend crucially on the burden sharing regime (OECD, 2012), a comparison of the regional distribution of mitigation costs is beyond the scope of this report.

This section further explores how such a stylised mitigation policy affects the economy until 2060, using the finer-grained ENV-Linkages model. Unfortunately, not enough information is available on the size of the various climate impacts for projections with lower levels of climate change that may result from stringent mitigation action. A more elaborate analysis of the benefits of action for stringent policy action requires substantially more data, for instance related to the climate impacts in a RCP2.6 type scenario with limited climate change (equivalent to 2 degrees global average temperature increase). Thus, it is impossible to robustly assess the benefits of a stringent mitigation scenario like the one investigated here. As an approximation, the analysis of avoided damages from mitigation presented here scales damages back from their baseline level using the difference in global temperature increase between both scenarios. In line with what e.g. Roson and Van der Mensbrugghe (2012) assume for health impacts, the ad hoc assumption is made that when no information is available on the level of damages for the RCP2.6, damages are scaled back in a way that assumes that only temperature changes above year 2000 levels lead to damages. For agricultural impacts, to account for the non-linearity in impacts, the time profile of the impacts are shifted such that a given temperature increase results in the same yield shocks, regardless of the speed of climate change. Thus, the slower temperature increases in the mitigation policy scenario leads to a postponement of the yield impacts. Effectively, the agricultural damages have thus been scaled by temperature rather than by time.

Many studies point to specific policy options that can be effective in reducing emissions at low costs. Prominent examples include the *New Climate Economy* report (GCEC, 2014), the OECD's own *Aligning policies for a low carbon transition* project (OECD, 2015d) and the IEA's *WEO Special Report on Energy and Climate change* (IEA, 2015). Such discussions are beyond the scope of this report, which limits itself to an evaluation of the avoided damages from mitigation action and a comparison of the sectoral and regions distribution of avoided damages and mitigation costs.

4.3.1. *The avoided damages from selected climate change impacts through mitigation policy action*

Section 4.2 already showed that implementing a stringent mitigation policy can reduce a large share of the potential damages from climate change, but not all. There are several reasons why damages will not go to zero: some climate impacts are already locked-in and at some stage the costs of further reducing damages outweigh the additional benefits. Panel A of Figure 4.8 shows to what extent the regional damages for 2060 as presented in Chapter 2 (cf. Figure 2.10) are avoided by a global price on carbon. At the global scale around half of the GDP costs of inaction (1% of GDP) are avoided by the mitigation scenario. The remaining damages amount to 1%. The mitigation policy can also reduce the uncertainties on economic losses, as illustrated in Panel B. The uncertainty on the remaining global damages ranges from 0.5% to 1.6% for the likely ECS range, and from 0.3% to 2.0% for the larger range (not shown in Figure 4.8). Thus, the band width of the global damages reduces from 2.3% of GDP (the difference between 1.0% and 3.3%) to 1.1%. This confirms the potential of mitigation action to reduce the risks of severe climate change: for the very high ECS value of 6 degrees, mitigation reduces the projected damages to GDP at the global level from 4.4% to 2.0%, and for e.g. India from 10.0% to 4.1%.

Not all regions will be equally affected by the changes in climate damages resulting from the mitigation policy. The composition of the various damage categories differs across regions, and not all categories are equally affected by the slower build-up of greenhouse gases in the atmosphere. Thus, even with the ambitious mitigation policy simulated here, there will be significant impacts of climate change in the more vulnerable regions. For the most severely affected regions, these projections show that the mitigation policy is more effective in reducing damages in regions with higher agricultural damages, such as India. Nonetheless, in most of Africa and Asia, the damages remain well above 1% of GDP.

For the OECD countries, where the costs of inaction are lower (cf. Figure 2.10 in Chapter 2), the avoided damages are logically also smaller. There are, however some striking differences: for instance, in the United States 46% of damages are avoided, while in Chile this percentage amounts to 54%, despite smaller gains in agriculture. This is primarily related to the (ad-hoc) assumptions that determine how impacts on health, energy and tourism scale back in presence of mitigation policies. Especially for tourism impacts, there is insufficient information to robustly assess this. The net benefits from climate change in Canada and Russia remain, albeit at a lower level under the stringent mitigation policy (and excluding the indirect effect of changes in energy demand on fuel exports as induced by mitigation). Again, the ad-hoc assumption on how tourism impacts scale with the level of climate change plays a crucial role in this result.

Panel B of Figure 4.8 shows the uncertainty on the regional avoided damages. These should be seen in conjunction with the uncertainties on the costs of inaction as presented in Panel B, Figure 2.10 in Chapter 2: high climate sensitivity implies high costs of inaction, but also high reductions in damages from mitigation. Avoided damages in India may be as high as 4.9% of GDP, but only if the damages in the no-mitigation projection amount to 8.4% of GDP. In other words, the uncertainties on avoided and remaining damages move together. In general, though, the percentage of avoided damages from mitigation action increases with the climate sensitivity. This is because the most severe impacts have the largest consequences on GDP, and limiting climate change thus has larger marginal effects.

Figure 4.8. **Regional damages from selected climate change impacts with and without mitigation policy**

Percentage change in GDP in 2060 w.r.t. no-damage baseline

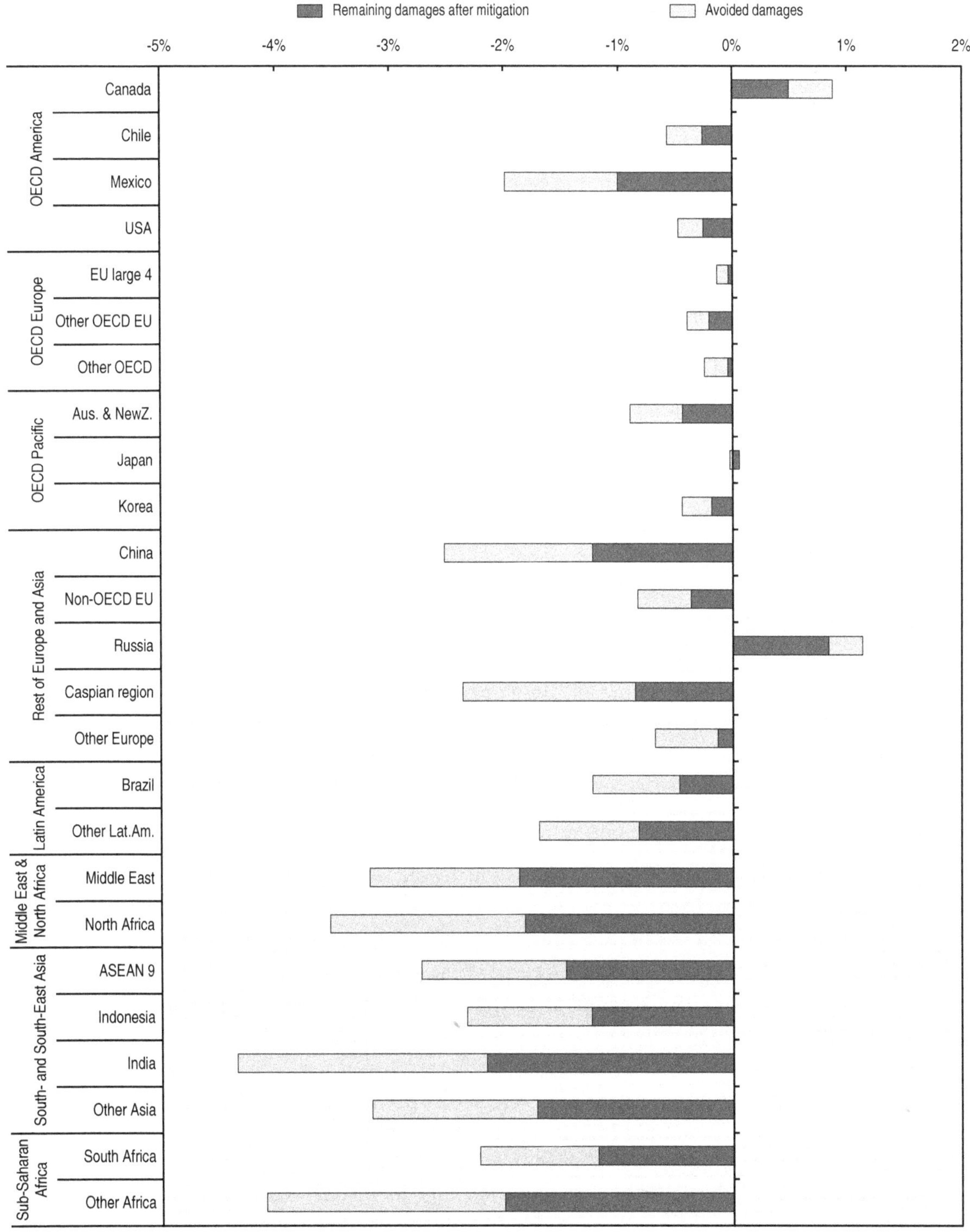

Figure 4.8. **Regional damages from selected climate change impacts with and without mitigation policy** *(cont.)*

B. Regional sensitivity of avoided damages to climate uncertainty

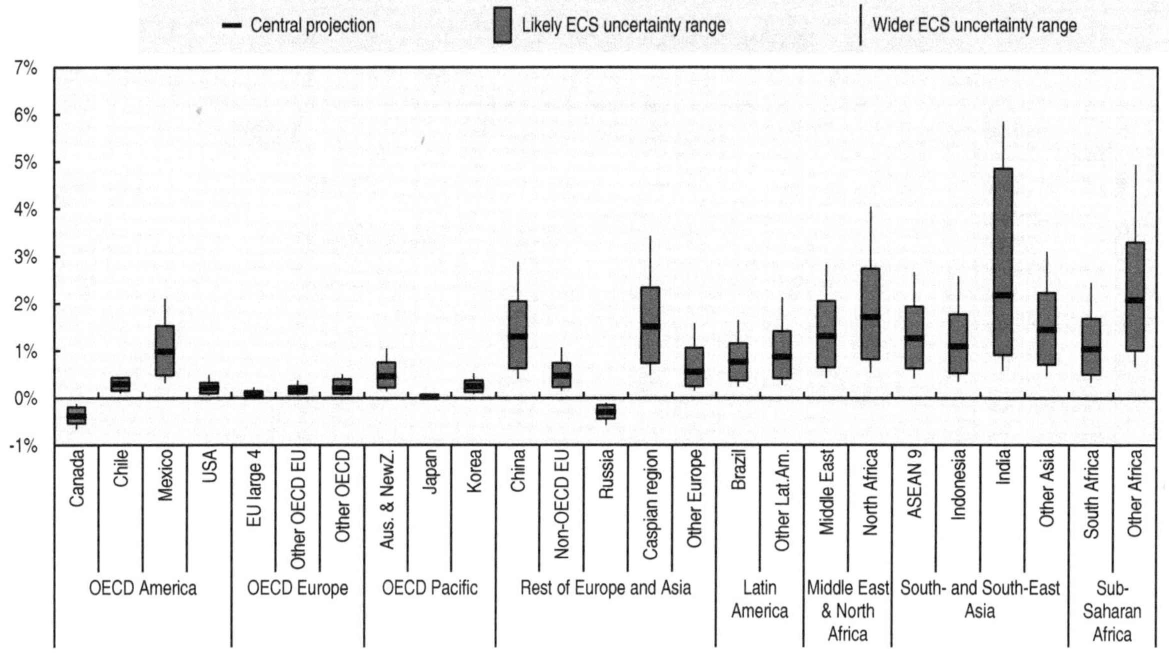

Source: ENV-Linkages model.

StatLink http://dx.doi.org/10.1787/888933276190

4.3.2. Sectoral distributional consequences of mitigation policy

To compare the impacts of climate change damages and the ambitious mitigation policy on the sectoral structure of the global economy, Figure 4.9 shows for 2060 the sectoral composition of the global GDP loss in Panel A. Panel B presents the associated projected change in the shares of different sectors in global GDP, with the no-damage baseline projection shares given with the labels on the x-axis. As these original shares are in percentage, their change is measured in percentage-points. For example, as illustrated in Figure 4.9, a 2%-point increase in the share of services in global GDP when climate change damages are considered implies that the share of these sectors is projected to increase from 58% to 60%.

Chapter 2 already found that climate damages tend to propagate throughout the economy and affect all sectors. Panel A clearly shows that all sectors contribute to the GDP loss, both from climate damages and from the mitigation policy. But the sectoral composition is strikingly different. This insight is reinforced by the changes in the share of specific sectors in total GDP as shown in Panel B.

The agricultural sector is negatively affected by both damages and mitigation efforts. Despite its relatively small size (projected at 2% of global GDP in 2060), the reduction in its share of GDP is significant. As explained in Chapter 2, there are substantial direct impacts on agriculture, and the GDP losses are relatively high in regions where agriculture forms a relatively large part of the economy. For the mitigation policy, the large emissions of methane and nitrous oxide, combined with CO_2 emissions from fuel combustion, imply that an economy-wide price on carbon leads to substantial cost increases in these sectors.

Figure 4.9. **Changes in structure of the global economy from damages from selected climate change impacts and mitigation policy**

A. Shares of sectors in global GDP loss in 2060 (percentage of total GDP loss)

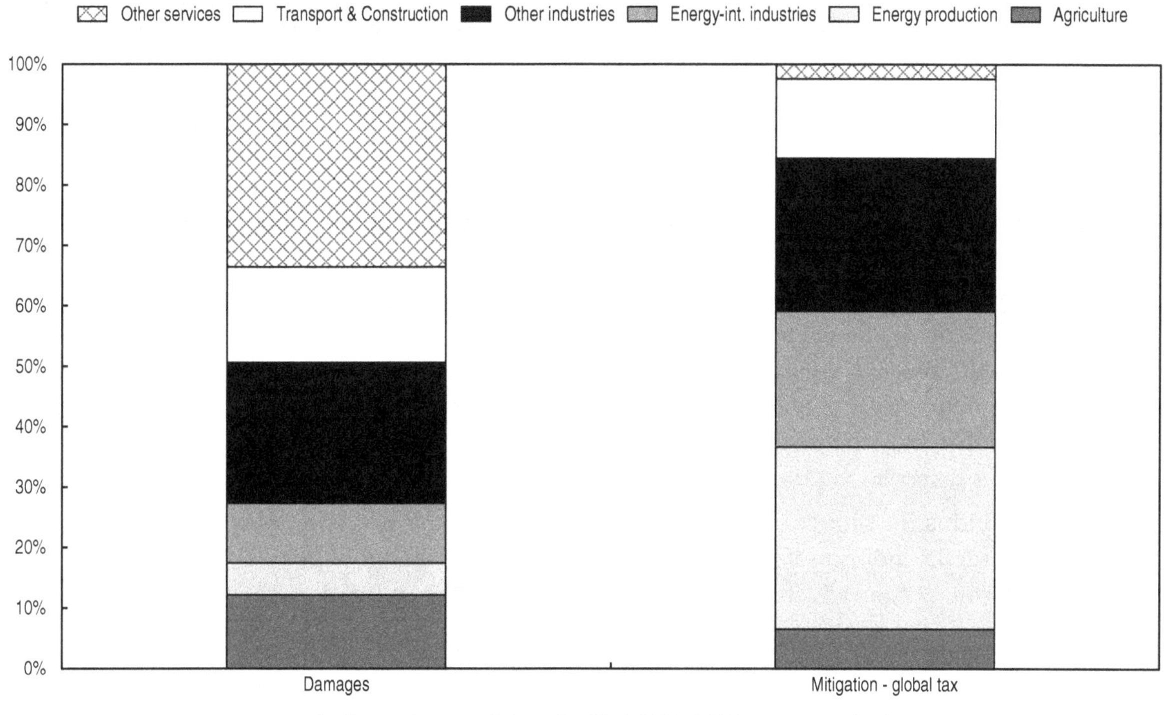

B. Changes in sectoral shares in global GDP in 2060 (percentage-points)

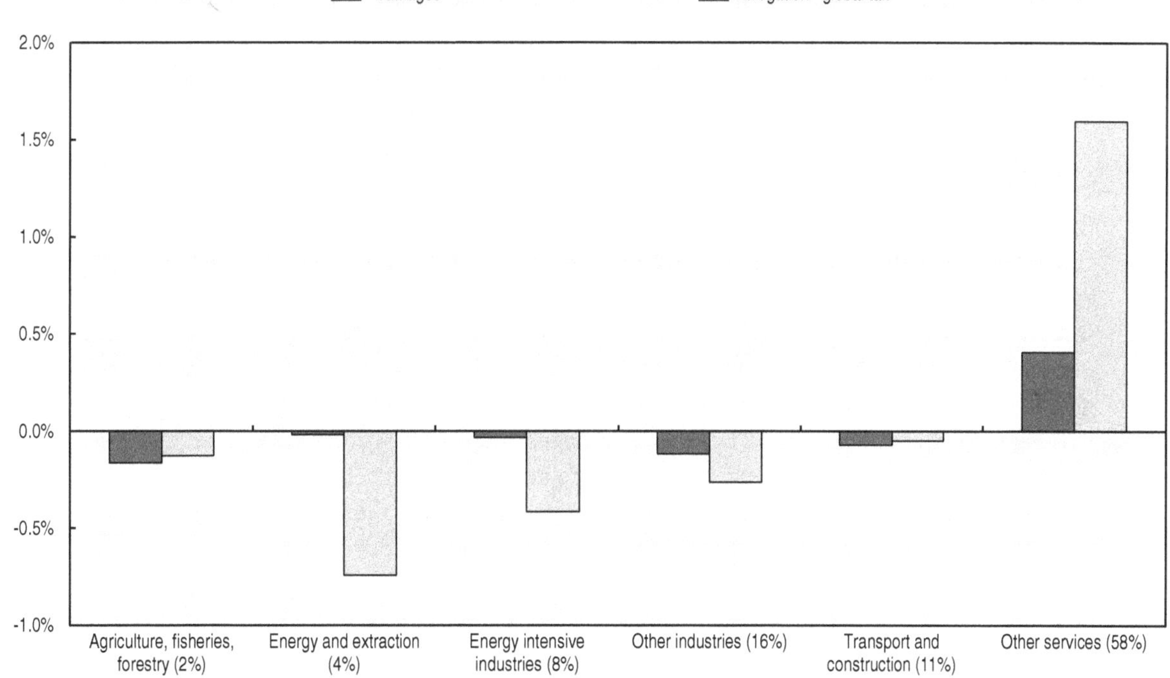

Source: ENV-Linkages model.

StatLink http://dx.doi.org/10.1787/888933276201

Like in other sectors, the availability of low-carbon technology options, as taken into account in the modelling exercise, limits these costs; with current technology, the costs would be substantially higher. The mitigation policy will, however, reduce the damages to agriculture, so it is still in the self-interest of the agricultural sector to ensure that global GHG emissions are reduced. These results also indicate how closely this sector is linked to the climate, through its dependence on weather and its contribution to GHG emissions, and hence to climate change.

The model simulations show a relatively small effect of climate damages (i.e. in absence of mitigation policies) on the energy and extraction sectors. To some extent, this is biased by the absence of damages directly on energy supply (due to lack of data on how climate impacts affect the various energy supply technologies). Nonetheless, it is clear that energy production and extraction, which are heavily emitting production activities, will be negatively affected by the mitigation policy. Behind this decline are much more diversified changes, however. The large decline comes fully for the fossil fuel producers. Cleaner energy technologies, including renewable power generation, can substantially increase their production activities. On balance, however, the global decline in energy demand, as part of the response to the carbon tax, leads to a retraction of the energy sector.

The industrial sectors are in a similar position: while the direct damages on this sector are limited, the indirect effects do imply a reduction in economic activity, but this pales in comparison to the reductions that are projected for the mitigation policy. This holds especially for the energy-intensive industries, where carbon pricing implies a substantial increase in costs and a reduction in output.

The projection results for transport and construction show a story that is more similar to that of agriculture, although less pronounced. Both climate damages and carbon taxation will negatively affect these sectors, but this hides more significant shifts at the lower sectoral level. For damages, the construction services are affected more than the transport sectors, not least because of the labour productivity losses from heat stress in construction. For the mitigation policy, the cleaner construction sector is less affected than transport.

Services are at the other end of the spectrum from the energy and industrial sectors: they can benefit from the mitigation policy as they are cleaner. As described in Chapter 2, the direct and indirect damages to the services sectors are significant; Panel A of Figure 4.9 confirms this. However, given the large size of services compared to the other sectors, Panel B shows that the relative share of the Services sectors can increase, i.e. they are *relatively* less affected than other sectors. Furthermore, as the regions with the largest services sectors are less severely affected by climate change impacts, their weight in the global economy increases (not shown in Figure 4.9), which further strengthens the increase in the sector share of services. Thus, both damages and the mitigation policy lead to a shift in the structure of the economy towards more services. However, as mitigation policies reduce damages, the gains in the services sector may be smaller than when mitigation policies are investigated without consideration of the benefits of policy action.

The relatively small effect of damages and mitigation on the overall structure of the global economy does not preclude much more substantive shifts within the various regions and sectors. For instance, the mitigation scenario induces a change in the composition of energy production away from fossil fuels towards renewables. But these are minor compared to the changes in the structure of the economy that happen for non-climate related reasons between now and 2060.

Notes

1. Also realising that the costs and effectiveness of policies are in themselves also uncertain.
2. Implicitly, the modelling analysis assumes some improvements in air quality control around the world in the no-damage baseline; this affects especially the calculation of the exogenous radiative forcing from short-lived climate forcers.
3. Another example is AD-WITCH, which applies the AD-DICE/AD-RICE methodology to the WITCH model (Agrawala et al., 2011).
4. As is typical in perfect foresight models, there is an immediate jump in the policy response to adjust optimally to future damages. This is e.g. visible in Figure 4.1 as a rapid build-up of the adaptation stock by 2015 to correct for the (fixed) suboptimal starting level of the adaptation stock.
5. Although one has to acknowledge that smooth economic models that are based on the concept of marginal responses to shocks are ill-suited to adequately assess the state of the economy that would reflect such extreme conditions.
6. Note that the AD-DICE model does not include a mechanism where early mitigation efforts lead to lower future costs through induced innovation. Bosetti et al. (2009) and Stern (2013) among others have pointed to the influence of this mechanism and show that induced innovation effects tend to lead to higher mitigation levels in the first few periods.
7. As the adaptation and damage specification are not identical, even for low temperature increases, the Weitzman specification in fact even leads to somewhat lower damages near the end of the century. This is a result of the recalibration of the damage function to the ENV-Linkages damage projections, which implies a slightly different policy response in both specifications, even at low temperatures.
8. As the trade-off between adaptation and mitigation depends on the inter-temporal effects of both policies, higher (lower) discount rates will imply a larger (smaller) weight on adaptation as compared to mitigation in the optimal mix.

References

Agrawala, S. et al. (2011), "Plan or react? Analysis of adaptation costs and benefits using Integrated Assessment Models", *Climate Change Economics*, Vol. 2(3), pp. 175-208.

Bollen, J. and C. Brink (2014), "Air pollution policy in Europe: Quantifying the interaction with greenhouse gases and climate change policies", *Energy Economics*, Vol. 46, pp. 202-215.

Bosetti, V. et al. (2009), "Optimal energy investment and R&D strategies to stabilize atmospheric greenhouse gas concentrations", *Resource and Energy Economics*, Vol. 31, pp. 123-137.

Braconier, H., G. Nicoletti and B. Westmore (2014), "Policy Challenges for the Next 50 Years", *OECD Economic Policy Papers*, No. 9, OECD Publishing, Paris, *http://dx.doi.org/10.1787/5jz18gs5fckf-en*.

De Bruin, K.C. (2014), "Documentation and calibration of the AD-RICE2012 and AD-DICE2013 models", *CERE Working Paper*, No. 2014-3, Umeå.

De Bruin, K.C., R. Dellink and R.S.J. Tol (2009), "AD-DICE: an implementation of adaptation in the DICE model", *Climatic Change*, Vol. 95, pp. 63-81.

De Bruin, K.C., R. Dellink and S. Agrawala (2009), "Economic Aspects of Adaptation to Climate Change: Integrated Assessment Modelling of Adaptation Costs and Benefits", *OECD Environment Working Papers*, No. 6, OECD Publishing, Paris, *http://dx.doi.org/10.1787/225282538105*.

Global Commission on the Economy and Climate (GCEC) (2014), *Better growth, better climate: The new climate economy report*, Washington, DC.

Gollier, C. (2007), "Comment intégrer le risque dans le calcul économique?", *Revue d'économie politique*, Vol. 117(2), pp.209-223.

IEA (2015), *World Energy Outlook Special Report on Energy and Climate Change*, IEA, Paris.

IPCC (2013), *Climate Change 2013: The Physical Science Basis*, Contribution of Working Group I to the Fifth Assessment Report of the Intergovernmental Panel on Climate Change [Stocker, T.F., D. Qin, G.-K. Plattner, M. Tignor, S.K. Allen, J. Boschung, A. Nauels, Y. Xia, V. Bex and P.M. Midgley (eds.)]. Cambridge University Press, Cambridge, United Kingdom and New York, NY, USA, 1535 pp.

Nemet, G.F., T. Holloway and P. Meier (2010), "Implications of incorporating air-quality co-benefits into climate change policymaking", *Environmental Research Letters*, Vol. 5(1).

Nordhaus, W.D. (2012), "Integrated Economic and Climate Modeling,", in: P. Dixon and D. Jorgenson (eds.), *Handbook of Computable General Equilibrium Modeling*, Elsevier.

Nordhaus, W.D. (2011), *Estimates of the social cost of carbon: Background and results from the RICE-2011 model*, New Haven, Conn. Cowles Foundation for Research in Economics, Yale Univ.

Nordhaus, W.D. (2007), *A question of balance*, Yale University Press, New Haven, United States.

Nordhaus, W.D. (1994), *Managing the Global Commons: The Economics of the Greenhouse Effect*, MIT Press, Cambridge, MA.

OECD (2015a), *Water Resources Allocation: Sharing Risks and Opportunities*, OECD Studies on Water, OECD Publishing, Paris, *http://dx.doi.org/10.1787/9789264229631-en.*

OECD (2015b), *NAEC synthesis report: New Approaches to Economic Challenges*, OECD report SG/NAEC(2015)1, Paris.

OECD (2015c), *Climate Change Risks and Adaptation: Linking Policy and Economics*, OECD Publishing, Paris, *http://dx.doi.org/10.1787/9789264234611-en.*

OECD (2015d), *Aligning Policies for the Transition to a Low-carbon Economy*, OECD report C(2015)50, Paris.

OECD (2014), "Recommendation of the Council on the Governance of Critical Risks", C/MIN(2014)8, *www.oecd.org/mcm/C-MIN%282014%298-ENG.pdf.*

OECD (2013), *Water Security for Better Lives*, OECD Studies on Water, OECD Publishing, Paris, *http://dx.doi.org/10.1787/9789264202405-en.*

OECD (2012), *OECD Environmental Outlook to 2050: The Consequences of Inaction*, OECD Publishing, Paris, *http://dx.doi.org/10.1787/9789264122246-en.*

OECD (2011), *Future Global Shocks: Improving Risk Governance*, OECD Publishing, Paris, *http://dx.doi.org/10.1787/9789264114586-en.*

OECD (2009), *The Economics of Climate Change Mitigation: Policies and Options for Global Action Beyond 2012*, OECD Publishing, Paris, *http://dx/doi.org/10.1787/9789264073616-en.*

Ramsey, F.P. (1928), "A mathematical theory of savings", *The Economic Journal*, Vol. 38, pp. 543-559.

Roson, R. and D. van der Mensbrugghe (2012), "Climate change and economic growth: Impacts and interactions", *International Journal of Sustainable Economy*, Vol. 4, pp. 270-285.

Stern, N. (2013), "The structure of economic modelling of the potential impacts of climate change: Grafting gross underestimation of risk onto already narrow science models", *Journal of Economic Literature*, Vol. 51(3), pp. 838-859.

Stern, N. (2007), *The Economics of Climate Change: The Stern Review*, Cambridge and New York: Cambridge University Press.

UK Treasury (2003), *The Green Book: Appraisal and evaluation in central government*, London.

UNEP (2014), *The Adaptation Gap Report 2014*, United Nations Environment Programme (UNEP), Nairobi.

Warszawski, L. et al. (2014), "The Inter-Sectoral Impact Model Intercomparison Project (ISI-MIP): Project framework", Proceedings of the National Academy of Sciences (PNAS), Vol. 111 (9), pp. 3228-3232.

Weitzman, M.L. (2012), "GHG targets as insurance against catastrophic climate damages", *Journal of Public Economic Theory*, Vol. 14(2), pp. 221-244.

ANNEX I

Description of the modelling tools

The economic modelling framework: ENV-Linkages

The OECD's in-house dynamic computable general equilibrium (CGE) model – ENV-Linkages – is used as the basis for the assessment of the economic consequences of climate impacts until 2060. The advantage of using a CGE framework to model climate impacts is that the sectoral details of the model can be exploited. Contrary to aggregated IAMs, where monetised impacts are directly subtracted from GDP, in a CGE model the various types of climate damages can be modelled as directly linked to the relevant sectors and economic activities.

ENV-Linkages is a multi-sectoral, multi-regional model that links economic activities to energy and environmental issues; Chateau et al. (2014) provide a description of the model. The model is calibrated for the period 2013-60 using the macroeconomic trends of the baseline scenario of the *OECD's Economic Outlook* (OECD, 2014). The ENV-Linkages model is the successor to the OECD GREEN model for environmental studies (Burniaux et al., 1992).

Production in ENV-Linkages is assumed to operate under cost minimisation with perfect markets and constant return to scale technology. The production technology is specified as nested Constant Elasticity of Substitution (CES) production functions in a branching hierarchy (cf. Figure A1.1). This structure is replicated for each output, while the parameterisation of the CES functions may differ across sectors. The nesting of the production function for the agricultural sectors is further re-arranged to reflect substitution between intensification (e.g. more fertiliser use) and extensification (more land use) of crop production; or between intensive and extensive livestock production. The structure of electricity production assumes that a representative electricity producer maximizes its profit by using the different available technologies to generate electricity using a CES specification with a large degree of substitution. The structure of non-fossil electricity technologies is similar to that of other sectors, except for a top nest combining a sector-specific resource with a sub-nest of all other inputs. This specification acts as a capacity constraint on the supply of the electricity technologies.

The model adopts a putty/semi-putty technology specification, where substitution possibilities among factors are assumed to be higher with new vintage capital than with old vintage capital. In the short run this ensures inertia in the economic system, with

Figure A1.1. **Production structure of a generic sector in ENV-Linkages**

Note: This generic structure does not apply to energy and agricultural sectors.
Source: ENV-Linkages model.

limited possibilities to substitute away from more expensive inputs, but in the longer run this implies relatively smooth adjustment of quantities to price changes. Capital accumulation is modelled as in the traditional Solow/Swan neo-classical growth model.

The energy bundle is of particular interest for analysis of climate change issues. Energy is a composite of fossil fuels and electricity. In turn, fossil fuel is a composite of coal and a bundle of the "other fossil fuels". At the lowest nest, the composite "other fossil fuels" commodity consists of crude oil, refined oil products and natural gas. The value of the substitution elasticities are chosen as to imply a higher degree of substitution among the other fuels than with electricity and coal.

Household consumption demand is the result of static maximization behaviour which is formally implemented as an "Extended Linear Expenditure System". A representative consumer in each region – who takes prices as given – optimally allocates disposal income among the full set of consumption commodities and savings. Saving is considered as a standard good in the utility function and does not rely on forward-looking behaviour by the consumer. The government in each region collects various kinds of taxes in order to finance government expenditures. Assuming fixed public savings (or deficits), the government budget is balanced through the adjustment of the income tax on consumer income. In each period, investment net-of-economic depreciation is equal to the sum of government savings, consumer savings and net capital flows from abroad.

International trade is based on a set of regional bilateral flows. The model adopts the Armington specification, assuming that domestic and imported products are not perfectly substitutable. Moreover, total imports are also imperfectly substitutable between regions of origin. Allocation of trade between partners then responds to relative prices at the equilibrium.

Market goods equilibria imply that, on the one side, the total production of any good or service is equal to the demand addressed to domestic producers plus exports; and, on the other side, the total demand is allocated between the demands (both final and intermediary) addressed to domestic producers and the import demand.

CO_2 emissions from combustion of energy are directly linked to the use of different fuels in production. Other greenhouse gas (GHG) emissions are linked to output in a way similar to Hyman et al. (2002). The following non-CO_2emission sources are considered: i) methane from rice cultivation, livestock production (enteric fermentation and manure management), fugitive methane emissions from coal mining, crude oil extraction, natural gas and services (landfills and water sewage); ii) nitrous oxide from crops (nitrogenous fertilizers), livestock (manure management), chemicals (non-combustion industrial processes) and services (landfills); iii) industrial gases (SF6, PFCs and HFCs) from chemicals industry (foams, adipic acid, solvents), aluminium, magnesium and semi-conductors production. Over time, there is, however, some relative decoupling of emissions from the underlying economic activity through autonomous technical progress, implying that emissions grow less rapidly than economic activity.

Emissions can be abated through three channels: i) reductions in emission intensity of economic activity; ii) changes in structure of the associated sectors away from the "dirty" input to cleaner inputs, and iii) changes in economic structure away from relatively emission-intensive sectors to cleaner sectors. The first channel, which is not available for emissions from combustion of fossil fuels, entails end-of-pipe measures that reduce emissions per unit of the relevant input. The second channel includes for instance substitution from fossil fuels to renewable in electricity production, or investing in more energy-efficient machinery (which is represented through higher capital inputs but lower energy inputs in production). An example of the third channel is a substitution from consumption of energy-intensive industrial goods to services. In the model, the choice between these three channels is endogenous and driven by the price on emissions.

ENV-Linkages is fully homogeneous in prices and only relative prices matter. All prices are expressed relative to the *numéraire* of the price system that is arbitrarily chosen as the index of OECD manufacturing exports prices. Each region runs a current account balance, which is fixed in terms of the *numéraire*. One important implication from this assumption in the context of this report is that real exchange rates immediately adjust to restore current account balance when countries start exporting/importing emission permits.

As ENV-Linkages is recursive-dynamic and does not incorporate forward-looking behaviour, price-induced changes in innovation patterns are not represented in the model. The model does, however, entail technological progress through an annual adjustment of the various productivity parameters in the model, including e.g. autonomous energy efficiency and labour productivity improvements. Furthermore, as production with new capital has a relatively large degree of flexibility in choice of inputs, existing technologies can diffuse to other firms. Thus, within the CGE framework, firms choose the least-cost combination of inputs, given the existing state of technology. The capital vintage structure also ensures that such flexibilities are large in the long run than in the short run.

The sectoral and regional aggregation of the model, as used in the analysis for this report, are given in Tables A1.2 and A1.2, respectively.

Table A1.1. **Sectoral aggregation of ENV-Linkages**

Agriculture		Manufacturing	
	Paddy rice		Paper and paper products
	Wheat and meslin		Chemicals
	Other grains		Non-metallic minerals
	Vegetables and fruits		Iron and steel
	Sugar cane and sugar beet		Metals n.e.s.
	Oil Seeds		Fabricated metal products
	Plant fibres		Food products
	Other crops		Other manufacturing
	Livestock		Motor vehicles
	Forestry		Electronic equipment
	Fisheries		Textiles
Natural resources and energy		Services	
	Coal		Land transport
	Crude oil		Air and water transport
	Gas extraction and distribution		Construction
	Other mining		Trade other services and dwellings
	Petroleum and coal products		Other services (government)
Electricity (7 technologies)[1]			

1. Fossil-Fuel based Electricity; Combustible renewable and waste based Electricity; Nuclear Electricity; Hydro and Geothermal; Solar and Wind; Coal Electricity with CCS; Gas Electricity with CCS.
Source: ENV-Linkages model.

Table A1.2. **Regional aggregation of ENV-Linkages**

Macro regions	ENV-Linkages countries and regions
OECD America	Canada
	Chile
	Mexico
	United States
OECD Europe	EU large 4 (France, Germany, Italy, United Kingdom)
	Other OECD EU (other OECD EU countries)
	Other OECD (Iceland, Norway, Switzerland, Turkey, Israel)
OECD Pacific	Oceania (Australia, New Zealand)
	Japan
	Korea

Table A1.2. **Regional aggregation of ENV-Linkages** *(cont.)*

Macro regions	ENV-Linkages countries and regions
Rest of Europe and Asia	China Non-OECD EU (non-OECD EU countries) Russian Federation Caspian region Other Europe (non-OECD, non-EU European countries)
Latin America	Brazil Other Lat. Am. (other Latin-American countries)
Middle East and North Africa	Middle-East North Africa
South and South-East Asia	India Indonesia ASEAN9 (other ASEAN countries) Other Asia (other developing Asian countries)
Sub-Saharan Africa	South Africa Other Africa (other African countries)

Source: ENV-Linkages model.

The climate modelling framework: AD-DICE

The results of the ENV-Linkages framework are complemented with results from stylised simulations with the AD-DICE model to examine longer-term impacts of climate change on economic growth, i.e. beyond 2060.[1]

In AD-DICE2013R economic production leads to GHG emissions but carbon dioxide (CO_2) emissions from fossil fuel combustion are the only endogenous sources (as shown in Chapter 2, CO_2 is the gas which has by far the largest contribution to climate change. The amount of CO_2 emissions per unit of output is assumed to decrease over time due to technological development. In turn CO_2 emissions increase the stock of CO_2 in the atmosphere, resulting in climate change. Though climate change includes a multitude of phenomena (such as changes in precipitation, changes in weather variability, increased extreme weather), it is represented by changes in atmospheric temperature in this model. Overall climate change negatively affects society and the economy through various different impacts. GDP impacts due to climate change are modelled as a percentage decrease in production as a function of mean atmospheric temperature change compared to 1900 levels. Investments in mitigation will reduce CO_2 emissions per unit of output at a cost, which decreases over time due to technological change. By adjustments to the economy (i.e. adaptation) initial climate change damages (gross damages) can be reduced to residual damages at a cost.

The model comprises of a single region, representing the globe. The economic and impact modules of the AD-DICE are calibrated by aggregated the regional specifications from the AD-RICE model, which divides the world into 12 regions.

AD-DICE2013R is a forward-looking Ramsey growth model, where global utility is maximized over the model horizon (given endowments). The model has time periods of 5 years and has a time horizon of 300 years. Utility is a function of consumption per capita discounted over time and over income per capita (richer generation's consumption creates less utility than poorer generation). The model finds the optimal balance of capital investments, mitigation investments, adaptation investments, adaptation costs and consumption to maximize utility.

By default, the climate change damage estimates in the AD-DICE model replicate the net damages of the DICE model. For this specific project, after calibrating AD-DICE2013R to the DICE2013R model, the damage and adaptation parameters in AD-DICE2013R have been recalibrated to be in line with the assessment of damages in ENV-Linkages. Logically, this implies that the same subset of climate impacts is covered. The AD-DICE model uses a stylized damage function: impacts are first valued and aggregated such that temperature increases lead to direct decreases in production and damages are then directly subtracted from GDP. A more detailed description of damages would include a production function approach, which includes the effects on production inputs and direct utility effects. There remains a large degree of uncertainty regarding the damages associated with climate change, where particularly many impacts have not yet been identified or quantified. The quantified damages in this model could be seen as a lower bound to expected climate change damages, but damages could be significantly higher than projected.

The AD-DICE model includes 2 forms of adaptation namely proactive adaptation, reactive adaptation. This distinction has been made to enable a more accurate description of the costs and benefits of different forms of adaptation and hence the total adaptation costs. Reactive adaptation describes adaptation measures that can be taken in reaction to climate change or climate change stimuli. This form of adaptation comes at a relatively low cost and is generally undertaken by individuals. Examples of this form of adaptation are the use of air-conditioning or the changing of crop planting times. Proactive adaptation on the other hand refers to adaptation measures that require investments long before the effects of climate change are felt. This form of adaptation usually requires large scale investments made by governments. Examples of this form of adaptation are research and development into new crop types or the construction of a dam for irrigation purposes.

The net damages of the DICE2013R models are separated into adaptation costs and residual damages. Firstly the gross damages (damages before/without adaptation) in period t are defined as follows:

$$GD_t = \alpha_1 \cdot T_t + \alpha_2 \cdot T_t^{\alpha_3}$$

where α_1 and α_2 are positive damage parameters, α_3 ranges between 1-4 and T represents the level of atmospheric temperature increase compared to 1900.

These are the damages that occur if no adaptation takes place, and are thus higher than the net damages. These damages can be reduced through the use of adaptation, assuming the following relationship:

$$RD_t = \frac{GD_t}{1 + P_t}$$

where P_t is the total level of protection (stock and flow) and RD_t are the residual damages. This functional form is chosen because it limits the fraction by which the gross damages can be reduced to the interval of 0 to 1. When total protection reaches infinity, all gross damages are reduced (the residual damages are zero) and when no protection is undertaken no gross damages are reduced (residual damages equal gross damages). This functional form also ensures decreasing marginal damage reduction of protection, that is the more protection is used the less effective additional protection will be. This is assumed as more effective, efficient measures of adaptation will first be applied whereas less effective measures after that.

Two forms of adaptation (stock and flow) together create total adaptation. The two forms of adaptation are aggregated together using a Constant Elasticity of Substitution (CES) function. Here the elasticity of substitution can be calibrated to reflect the observed relationship between the two forms. This function is given as follows:

$$P_t = \gamma \cdot (v_1 SAD_t^{\rho_A} + v_2 FAD_t^{\rho_A})^{v_3/\rho_A}$$

where SAD_t is the total amount of adaptation capital stock. FAD_t is the amount spent on flow adaptation in that period. Furthermore, $\rho_A = \frac{\sigma - 1}{\sigma}$, where σ is the elasticity of substitution.

Adaptation capital stock is built up as follows:

$$SAD_{t+1} = (1 - \delta_k) SAD_t + IAD_t$$

where δ_k is the depreciation rate and IAD_t are the investments in stock adaptation (SAD_t).

The adaptation module of AD-DICE2013R is calibrated based on estimates of adaptation costs and benefits from the impact literature. More precisely, based on the various impact estimates discussed in Tol (2009), the relative importance of each impact sector per region is estimated. Furthermore, for each climate impact sector the adaptation costs and benefits for each region were estimated based on available impact studies and expert judgment. Finally, the regional adaptation costs and benefits were aggregated to a global level. For a full description of this process, please refer to De Bruin (2014).

Climate change is global environmental problem, affecting all regions of the world both now and in centuries to come. Both the causes of climate change (different sources of GHG emissions) and the effects of climate change are innumerable, diverse, and vary is scope and scale. Attempting to include all causes and effects of climate change in a single model is a difficult task. Especially estimating the effects of climate change in the long run is a complex process, which involves many uncertainties. IAMs are tools created to assess the effects of the economy on climate change and vice versa in the long run. Due to the many mechanisms involved and the long time frame, these models need to make (simplifying) assumptions. IAMs are hence highly aggregated top-down models, which do not include all sectoral and regional impacts in detail. Though these assumptions and simplifications are necessary due to both lack of data (it is hard to predict future effects) and computational limitations, they do form a significant drawback of IAMs. Given these drawbacks applying a model such as AD-DICE can still give important insights into the magnitude and development of both the economy and the climate. Given that climate change is both a global problem and will have the greatest affects in the long term, an analysis of climate change is incomplete without a global long-term perspective. The strength of IAMs such as AD-DICE is that they can shed some light on the long-term climate consequences of our actions now.

The *Weitzman damage* specification is used as a sensitivity analysis for the damages assumed in the model. Weitzman (2012) proposes a damage function which increases much more steeply over temperature change than the DICE/AD-DICE damage function. The Weitzman function is similar to the original DICE/AD-DICE damage function for low temperatures but is significantly higher for higher temperature change levels.[2] The DICE damage function assumes that net damages as a fraction of GDP consist of a linear component and a quadratic component, as follows:

$$NHD_t = \alpha_{NH1} \cdot T_t + \alpha_{NH2} \cdot T_t^2$$

where $\alpha_{NH1}, \alpha_{NH2}$ are damage parameters.

The Weitzman damage function has the following form:

$$WD_t = \left(\frac{T_t}{20.46}\right)^2 + \left(\frac{T_t}{6.08}\right)^{6.76}$$

The AD-DICE Weitzman damage function is calibrated in AD-DICE2013R to replicate the same level of net damages as the original Weitzman damage function. Furthermore, adaptation costs and benefits are assumed to match the original AD-DICE damage specification for damages that are in line with the DICE damage function, i.e. for low temperature increases. Any additional damages in excess of the DICE damage function that are projected by the Weitzman function are assumed to be catastrophic damages (see de Bruin (forthcoming) for details). As it is estimated that the potential of adaptation to reduce catastrophic damages is limited, adaptation plays a much smaller role with the Weitzman damages specification compared to the original specification, particularly in the long run. Consequently, both specifications deliver very similar results at low temperature levels, but the Weitzman specification projects much larger damages and lower adaptation levels for higher temperature increases.

The calibration of AD-DICE2013R for the CIRCLE project is carried out in several steps. First, an AD-DICE2013R replication of the DICE2013R model is given. Then, the socioeconomic baseline and damage projections until 2060 are taken from the ENV-Linkages model. This implies a change in the parameters for population growth, economic activity, technological progress, and the parameters in the damage and adaptation functions. The revised adaptation parameters take the sectoral composition of damages in ENV-Linkages into account, i.e. the costs and effectiveness of adaptation is adjusted to reflect the share of the various impacts in ENV-Linkages. Finally, the central projection contains a number of choices that are not standard in the original DICE model: an equilibrium climate sensitivity of 3.0, discount factors based on the UK Treasury rules, and – not least – only flow adaptation measures are considered.

The resulting central projection in AD-DICE2013R can be linked to the damage projections of the original DICE model in the following way:

	2060 %	2100 %
1. Central projection	**2.1**	**5.8**
2. 1 plus full adaptation (instead of flow adaptation only)	1.8	4.6
3. 2 plus Nordhaus discounting instead of UK Treasury discounting	1.8	4.6
4. 3 plus a equilibrium climate sensitivity of 2.9 instead of 3.0	1.7	4.4
5. 4 plus DICE damages instead of ENV-Linkages damages until 2060	1.6	4.3
6. 5 plus DICE socioeconomic baseline instead of CIRCLE baseline	1.5	4.0
7: original DICE: 6 without explicit adaptation	1.5	4.0

As the table shows, the main difference between the AD-DICE central projection and the original DICE projection comes from the assumption in AD-DICE that not all adaptation options are available in absence of policies (approximated by excluding stock adaptation). The socioeconomic baseline of CIRCLE, with somewhat higher emissions, and the higher climate sensitivity in the CIRCLE specification also contribute to the difference: these lead to higher temperature increases and hence larger damages. The other adjustments, including the recalibration to the ENV-Linkages damage levels have much smaller impacts on the results.

Notes

1. AD-DICE does not allow calibrating climate change damages using the detailed production function approach detailed in Section 1.3.
2. Note that the Weitzman damage function is calibrated to the DICE replication in AD-DICE, and has not been recalibrated to the CIRCLE specification.

References

Burniaux, J.-M., G. Nicoletti and J. Oliveira Martins (1992), "GREEN: A Global Model for Quantifying the Costs of Policies to Curb CO_2 Emissions", *OECD Economic Studies*, 19 (Winter).

Chateau, J., R. Dellink and E. Lanzi (2014), "An Overview of the OECD ENV-Linkages Model: Version 3", *OECD Environment Working Papers*, No. 65, OECD Publishing, Paris, *http://dx.doi.org/10.1787/5jz2qck2b2vd-en*.

De Bruin, K.C. (2014), "Documentation and calibration of the AD-RICE2012 and AD-DICE2013 models", *CERE Working Paper*, No. 2014-3, Umeå.

Hyman, R.C., J.M. Reilly, M.H. Babiker, A. De Masin and H.D. Jacoby (2002), "Modeling Non-CO_2 Greenhouse Gas Abatement", *Environmental Modeling and Assessment*, Vol. 8, No. 3, pp. 175-86.

IPCC (2007), *Climate Change 2007: Impacts, Adaptation and Vulnerability*, Contribution of Working Group II to the Fourth Assessment Report of the Intergovernmental Panel on Climate Change, M.L. Parry, O.F. Canziani, J.P. Palutikof, P.J. van der Linden and C.E. Hanson, Eds., Cambridge University Press, Cambridge, UK, 976pp.

Nordhaus, W.D. (2012), "Integrated Economic and Climate Modeling,", in: P. Dixon and D. Jorgenson (eds.), *Handbook of Computable General Equilibrium Modeling*, Elsevier.

OECD (2014), *OECD Economic Outlook*, Vol. 2014/1, OECD Publishing, Paris, *http://dx.doi.org/10.1787/eco_outlook-v2014-1-en*.

Tol, R.S.J. (2009), "The Economic Effects of Climate Change", *Journal of Economic perspectives*, Vol. 23(2), pp. 29-51.

Weitzman, M.L. (2012), "GHG targets as insurance against catastrophic climate damages", *Journal of Public Economic Theory*, Vol. 14(2), pp. 221-244.

ORGANISATION FOR ECONOMIC CO-OPERATION AND DEVELOPMENT

The OECD is a unique forum where governments work together to address the economic, social and environmental challenges of globalisation. The OECD is also at the forefront of efforts to understand and to help governments respond to new developments and concerns, such as corporate governance, the information economy and the challenges of an ageing population. The Organisation provides a setting where governments can compare policy experiences, seek answers to common problems, identify good practice and work to co-ordinate domestic and international policies.

The OECD member countries are: Australia, Austria, Belgium, Canada, Chile, the Czech Republic, Denmark, Estonia, Finland, France, Germany, Greece, Hungary, Iceland, Ireland, Israel, Italy, Japan, Korea, Luxembourg, Mexico, the Netherlands, New Zealand, Norway, Poland, Portugal, the Slovak Republic, Slovenia, Spain, Sweden, Switzerland, Turkey, the United Kingdom and the United States. The European Commission takes part in the work of the OECD.

OECD Publishing disseminates widely the results of the Organisation's statistics gathering and research on economic, social and environmental issues, as well as the conventions, guidelines and standards agreed by its members.

OECD PUBLISHING, 2, rue André-Pascal, 75775 PARIS CEDEX 16
(97 2015 10 1 P) ISBN 978-92-64-23540-3 – 2015

www.ingramcontent.com/pod-product-compliance
Lightning Source LLC
LaVergne TN
LVHW081407110826
845149LV00010B/1662

* 9 7 8 9 2 6 4 2 3 5 4 0 3 *